Insect Architecture

Insect Architecture

How Insects Build, Engineer, and Shape their World

TOM JACKSON

Consultant editor:
MICHAEL S. ENGEL

PRINCETON UNIVERSITY PRESS
PRINCETON AND OXFORD

Published in 2025 by Princeton University Press
41 William Street, Princeton, New Jersey 08540
99 Banbury Road, Oxford OX2 6JX
press.princeton.edu

Copyright © 2025 by Quarto Publishing plc

GPSR Authorized Representative: Easy Access System Europe - Mustamäe tee 50, 10621 Tallinn, Estonia, gpsr.requests@easproject.com

All rights reserved, Princeton University Press is committed to the protection of copyright and the intellectual property our authors entrust to us. Copyright promotes the progress and integrity of knowledge created by humans. Thank you for supporting free speech and the global exchange of ideas by purchasing an authorized edition of this book. If you wish to reproduce or distribute any part of it in any form, please obtain permission.

Requests for permission to reproduce material from this work should be sent to permissions@press.princeton.edu

Ingestion of any PUP IP for any AI purposes is strictly prohibited.

ISBN: 978-0-691-27523-9
Ebook ISBN: 978-0-691-27524-6
Library of Congress Control Number: 2025933320
British Library Cataloging-in-Publication Data is available

This book was conceived, designed, and produced by
The Bright Press, an imprint of The Quarto Group
1 Triptych Place
London SE1 9SH
United Kingdom
www.quarto.com

The Bright Press
Publisher: James Evans
Editorial Director: Isheeta Mustafi
Managing Editor: Lucy Tipton
Publishing Operations Director: Kathy Turtle
Publishing Assistant: Jemima Solley
Art Director: James Lawrence
Editor: Nick Pierce
Project Editor: Katie Crous
Layout and Image Research: Lindsey Johns
Illustration: Sandra Pond
Production Controller: George Li

Front cover photo: Shutterstock/2021 Photography
Back cover photo: Shutterstock/Mrs. Nuch Sribuanoy

Printed in Malaysia

10 9 8 7 6 5 4 3 2 1

Contents

	Foreword	6
	The Evolution of Insects	8
	Introduction	10
CHAPTER ONE	**Beetles and Bugs**	16
CHAPTER TWO	**Web Spinners and Silk Weavers**	40
CHAPTER THREE	**Funnels, Cases, and Stalk Builders**	64
CHAPTER FOUR	**Wasps**	82
CHAPTER FIVE	**Bees**	102
CHAPTER SIX	**Ants**	126
CHAPTER SEVEN	**Termites**	146
	Afterword	164
	Glossary	168
	Index	171
	Acknowledgments and Picture Credits	176

Foreword
A Love for Building

There is an irresistible desire to build. Whether this comprises a rough accumulation of stones forming boundaries in ancient caves or the vast steel girders that reach upward as if trying to grasp the sky, we love to build. Individual homes extend into complex communities and from there to towns, cities, and sprawling megalopolises with 10 million or more individuals pulsing through them like blood in veins. Our constructions are both functional and aesthetic, and since antiquity we have highlighted the greatest of architectural achievements as wonders of the world—typically allocated in septets.

Architecture can be characterized as anything that is purpose-built, including both the bringing together of exogenous or endogenous materials, or the excavation of negative spaces (for example, tunnels). A simple burrow dug into the ground for the purpose of building a protective chamber for a developing brood and food storage can be recognized as architecture at, perhaps, its simplest, whereas the simple movement of earthworms through the soil and leaving behind a tunnel as a byproduct is not. As we know, however, architecture is so much more than these simplest of examples, as we watch birds build nests, voles dig chambers and line them with plant material or even collected feathers, beavers build dams—the world is filled with diverse animals building diverse structures.

Our vast urban landscapes are home not only to us but to many other species, the most abundant of which are the insects. Whether we like it or not, insects are an omnipresent and beneficial element of our lives, and they are an inextricable component of our manufactured worlds. We are more like insects than we perhaps wish to think. Our cities operate much like large insect societies, and insects share with us that desire to build. Insects have their manufactured homes, neighborhoods, skyscrapers, and cathedrals, and these were crafted and refined long before the earliest of primates stood upon this Earth.

Indeed, for tens of millions of years insects have been building familiar structures, ranging from simple delimited borders for a brooding chamber to intricate cities housing millions, and many of these with labyrinthine chambers for climate control, waste disposal, and even subterranean gardens. Today's architectural design and engineering rightly looks to the insects and their eons of experience for insights into how to build smarter and more efficient buildings, and there are any number of modern marvels, with their rounded surfaces and cavities that evoke life in an ant or termite mound or even the hexagonal hives of bees. In one way or another insects did it first, therefore it is only right that we should, as the poet of Proverbs said, "Go to the ant, thou sluggard; consider her ways, and be wise." Insects can teach us much, and their architecture is as awe-inspiring as ours.

Michael S. Engel

RIGHT
SILKEN SHELTER
The egg-shaped cocoons of silk moths (Bombyx mori) are woven from a single strand of silk that might be more than half a mile long.

BELOW
MUD DAUBER
A member of the Eumenidae applies a blob of mud to a nest constructed for a single larva.

The Evolution of Insects

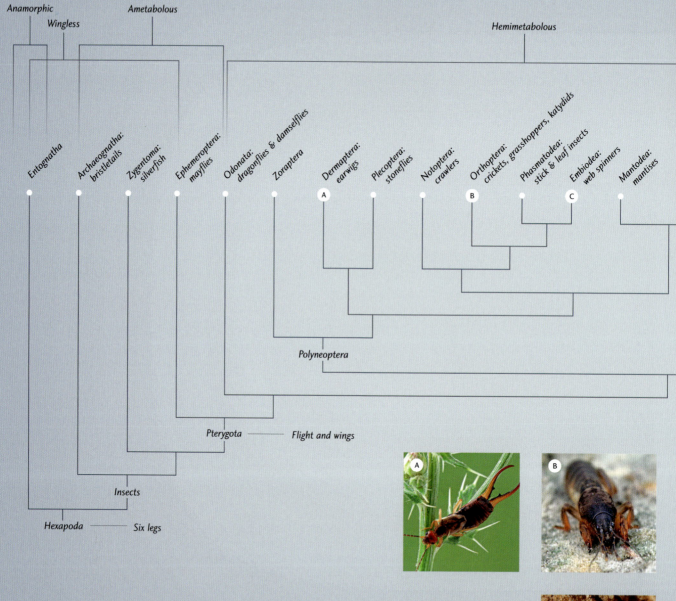

The chart above shows the evolutionary relationships between the insect orders (including some that are not mentioned in this book), along with some non-insect hexapods. Among other things, it illustrates how insect architecture is not the preserve of a small subset of insects, but found across the whole class of animals.

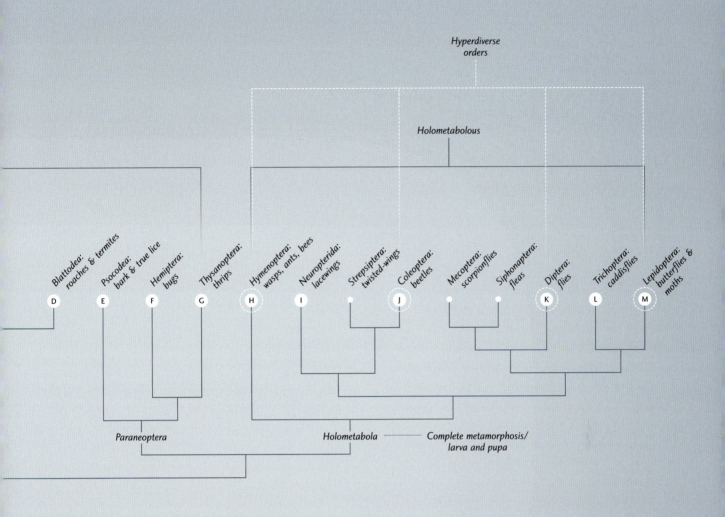

Introduction

From beavers to bowerbirds, trapdoor spiders to swallows, humans are merely the latest in a long line of animals that build and maintain elaborate structures. Just like us, these animal architects are making a place to live, rest, and raise young. The animals that have developed the widest range of building skills, however, are undoubtedly the insects. Insects create vast earthworks or dig underground tunnel networks. They make paper walls, stitch folded leaves, glue together pebbles, and also build with wax, mud, and even their own feces.

While faced with broadly the same structural challenges as humans, insects have come up with some very different solutions. Human engineers are continually developing new materials inspired by insect innovations, and human architects are to this day using insect designs to improve the form, function, and efficiency of buildings. For example, honeycomb structures are routinely used to reduce weight, just as they do in a bee hive; the ventilation system of large buildings is informed by the way air flows in and out of termite mounds to maintain stable conditions inside; and iterative algorithms, where construction processes are modified and repeated, mimicking the way insects build their structures, are used to design innovative structures and appealing building complexes.

WHAT ARE INSECTS?

Insects are small creatures, generally measured in millimeters. Taking the different sizes into account helps to appreciate the sheer scale of insect buildings. If an average cathedral termite mound were to be scaled up to human size, it would be taller than anything built by humans. The underground nest of a leaf-cutter ant would be the size of Paris. There is no need for nightmares, however—these little bugs and beasties cannot grow to human size because they are limited by their physiology.

Insects belong to a group of animals called the Arthropoda. This name means "jointed legs" and is self-explanatory enough. Any animal with an exoskeleton and legs made up of several jointed sections belongs in this group. Along with the insects, then, this includes the spiders, scorpions, crustaceans, and millipedes (plus a few more). The defining feature of an insect is that the adult has six legs connected to a midbody section, or thorax, which sits between a head section and abdomen. Many insects also have one or two pairs of wings—but not all of them fly.

OPPOSITE
TERMITE TOWER
Termites (above) build big structures such as this cathedral termite mound in Litchfield National Park, Northern Territory, Australia, which is about 13 ft (4 m) high.

Essentially, this body plan is what sets them apart from the other arthropods, and it has proved to be a very effective one. The insects make up 90 percent of animal species, and half of all known living organisms are insects. (There are a lot of unknown living organisms, so that second figure will probably reduce over time.)

Given their dominance in numbers, the insects have diversified to live in all parts of the globe—although they are almost entirely excluded from marine habitats. The list of insects is familiar enough, but also inevitably incomplete: flies, wasps, ants, bees, butterflies, crickets, and beetles to name but a few among thousands more. They all share the body plan, but use it in very different ways.

The insects, as with all arthropods, have an external skeleton, or exoskeleton, rather than an internal one as we are more familiar with. The outer covering of the body is made from a hard but flexible material called chitin. This exoskeleton cannot grow along with the rest of the body inside, so when it becomes too tight, the skin is shed, or molted.

One of the most successful features of the insects is their mouthparts. The oral opening is surrounded by limb-like mouthparts, and these parts have been adapted to bite, suck, chew, slurp, and scrape. As we shall see, the mouth is the primary construction tool in building insect structures.

One thing the insect mouth does not do is breathe in air. Insects have no lungs. Oxygen arrives via many small tubes, called tracheae, which penetrate the exoskeleton. Generally speaking, gas exchange relies on passive diffusion. The slow rate of this kind of movement limits the length of the tracheae and thus limits the size of insects. There can be no dog-sized ants, bear-sized beetles, or wasps as big as eagles.

DEVELOPMENT SYSTEMS

Among the conformity of the insect body plan (and its incredible adaptability), there is a great schism in insectkind. The divide is to do with how the insects develop from the egg. Many of the most diverse and abundant orders, such as beetles, flies, moths, and wasps, are holometabolous. Most of the rest, which include bugs, such as aphids and cicadas, grasshoppers, roaches and termites, are hemimetabolous.

Hemi- means "half" or "part," *holo-* means "total," while *-metabolous* refers to "changeable." In hemimetabolous insects, the young form that emerges from the egg is called a nymph. It is a small wingless form of the adult insect, with a body made from three clear body sections, six legs, and a pair of antennae. As the nymph grows, molting several times, it will develop more adult features. In the final stage, it becomes a sexually mature adult (with wings if that species can fly). While it shares the adult body plan, the nymph can also share the same sources of food. The older and younger generations of hemimetabolous insects may, therefore, compete for resources.

Hemimetabolous insects go through a partial change as they develop; holometabolous species undergo a wholesale metamorphosis. The young form that emerges is a larva, which has a wormlike body. Larvae go by many names, such as caterpillars, maggots, and grubs. An insect larva is a feeding machine, a set of mouthparts that can wriggle around over, in, and through its food supply. Once it grows large enough, the larva will become inactive. Its skin thickens to become a capsule-like pupa. In many species, the pupa is housed inside a cocoon of woven silk or materials collected from the surroundings. Inside the pupal case, larval tissue is reorganized into an adult form.

OPPOSITE
TREETOP HOME
A team of weaver ants works to heave together the edges of a folded leaf. The edges will then be stitched together with sticky silk to form part of a nest.

Holometabolous development evolved in part because there were too few resources available in the egg. Therefore, the young hatches from the egg at an earlier stage, in a more embryonic, wormlike form. This larva then completes its development after locating a fresh source of food—a source that was not always available to its parents.

This generational divide has driven much of the evolution of insect architecture, as parents seek to provide their larvae with the best start in life. Additionally, the great gulf between the larval and adult forms can mean there are more building techniques and materials available. For example, it is common for insect larvae to produce silk, always useful in construction, but it is less commonly used by adults.

BUILDING PROCESSES

Not all insects are builders; in fact the majority are not. This book focuses largely on the insects that build more than just holes in the ground or wood, but rather construct something, however simple or complex, with a particular purpose.

A human architect is the person that designs a building and oversees its construction. An insect architect does it very differently. For one, there is no pre-planned design—just the drive to build. There is also no single overseer of large-scale projects, but instead a reactive system where the builders change their behaviors to suit the conditions. There is obviously a great deal of overlap, but five processes used by insect architects have been proposed, each with varying complexity and degrees of cooperation.

The first method is a fixed set of actions carried out in a precise order. This is the process used by web spinners (Chapter Two). The second process is used by solitary architects. It involves feedback loops where the next stage of construction is influenced by the results of the last.

ABOVE
NEW ARRIVAL
A newly emerged bee is tended to by a sister worker as it prepares to climb out of its wax cell on the brood comb for the first time.

OPPOSITE
HORNET'S NEST
The gallery of sturdy hexagonal brood cells can be seen through the bottom aperture of a hornet's nest encased in paper walls.

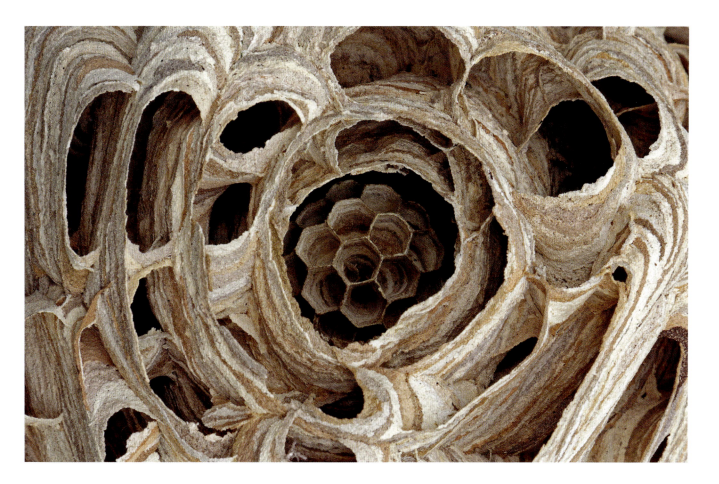

This is the system most commonly used by insects. It can result in precise constructions but is limited to small structures. Antlions (Chapter Three) dig their deadly pits this way and dung beetles (Chapter One) use it to entomb balls of dung.

The third method is one used by insects that build collectively. (In truth, each builder is still working alone, but they respond to the results of the group effort.) Harvester ants (Chapter Six) build a pebble-dashed mound above their underground nest using this method. The fourth building system involves cooperative building. The builders are a coherent team that communicate with one another to guide the construction. This is quite an unusual way of working. Weaver ants, which build nests from leaves, are one example of insect that work in this way.

The fifth and final process is the most complex, and it is used by wasps, bees, ants, and termites (Chapters Four to Seven) to construct their impressive nests. The building is coordinated, but, unlike the cooperative system of building just described, the builders are responding to stimuli from the structure itself that compel them to dig or to deposit, to demolish or repair. The stimuli are not fully understood, but undoubtedly involve chemical cues produced by workers and queens. For example, workers recruited to dig will produce a chemical that stimulates them to switch from digging to carrying the loose soil to a deposition pile elsewhere. The level of this deposit signal is at first low, but after a period of digging it will inevitably dominate, and so the workers change jobs and clear the spoil.

WHAT IS INSECT ARCHITECTURE?

This book casts its net widely when considering the wonders of insect architecture. As well as the insect cities built by some wasps, ants, and termites, it also introduces smaller settlements, which include the nests, traps, and nurseries constructed by a much wider range of insects. An underground lair is fine-tuned to amplify mating calls; hammocks of slimy snares glow in the dark; and a summer home is built from nothing but bubbles.

CHAPTER ONE
Beetles and Bugs

The words "beetle" and "bug" are often used synonymously, but they represent two very different orders of insects. Despite some very superficial at-a-glance similarities, beetles (Coleoptera) and bugs (Hemiptera) are far from the same. They display different life cycles and target largely different food sources. However, despite occupying different zones of the insect spectrum, the structures built by bugs and beetles are often burrows and tunnels, constructed beneath the soil, under rocks, or in wood. They are not alone in being good tunnelers. Earwigs (Dermaptera) and ice crawlers (Grylloblattodea) also create homes, however temporary, by digging. Additionally, mole crickets (Gryllotalpidae) are a standout group among the crickets and grasshoppers (Orthoptera). As their name alludes to, mole crickets are avid tunnelers that spend much of their lives underground. Not only are they adapted to this fossorial, or digging, lifestyle, the crickets use their building skills to construct amplifiers for their communication calls.

DIVIDING LINE

The insects covered in this chapter fall into two distinct groups: beetles and everything else. While the division might appear to be unbalanced, with one order set against four, beetles actually comprise at least three times as many species as the other four orders put together. Beetles are thought to be the biggest order of insects, with 350,000 species.

Bugs are a big group—there are around 80,000 species. Earwigs are less diverse, with fewer than 2,000 types. There are only around 100 mole cricket species (although there are thousands more in the wider cricket, locust, and grasshopper group). Meanwhile, ice crawlers are a very small order, numbering just 30 species, which reflects their very unusual adaptation to cold conditions.

What sets these orders apart is the way their young develop from the egg. Bugs, earwigs, and the others are hemimetabolous, and start life as nymphs, or mini-versions of the adult. The beetles, however, are holometabolous. Their chubby, wormlike larvae are generally termed "grubs." It is not uncommon to find grubs digging inside or among their food source. They excavate plant buds or fruits, they chew through wood or wriggle through soils to find roots to eat. Such hungry grubs can leave a trail of destruction, creating tunnels and holes, but these are not really structures built with a deliberate purpose. However, some beetles have become architects that build more elaborate constructions.

OPPOSITE TOP
FECAL FOOD
The grub of a dung beetle consumes the fecal matter of a larger herbivore. This food source retains 50 percent of its original nutritional value.

OPPOSITE BOTTOM
SPRUCE BARK BEETLE
A small beetle gnaws its way into the bark of a tree to build a tunnel nursery for its young.

SUCKERS

Bugs—true bugs in the entomological sense, that is—are characterized by a spike-like mouthpart used for sucking. While some bugs suck blood, most attack plants and drink their watery sap. Sap is a plentiful food source—although one very low in nutrients—and so there is enough to be shared between young and old. Nevertheless, one of the largest kinds of bug, cicadas, spend their nymphal days—often years—out of sight underground sucking sap from roots. The only aboveground evidence they are there are little turrets that work as air vents in the dry—and snorkels when it is wet! Meanwhile, some earwig species are among the insect world's most caring parents. They construct a safe nursery chamber for their eggs, carefully brooding them to maximize their chance of hatching. The mother keeps her newly hatched young at home, feeding them on her own partially digested regurgitated food. Now let's take a closer look at how beetles, bugs, and others create their homes.

BEETLES AND BUGS

BLUEPRINTS
Wax and Foam

Plant-feeding bugs invariably sustain themselves out in the open, in full view of predators and parasites, and in the glare of the sun. There is nowhere to hide, little option to move, and few ways of fighting back. Instead, bugs called scale insects (Coccoidea) construct homes armored with waxy resins. Meanwhile, froghoppers (Cercopoidea) hide away in a house made of bubbles.

FIG. 1
SCALE INSECTS' WAX DOME

Scale insects start out as nymphs, known as crawlers. They crawl around the branch looking for a suitable site to latch on to the host plant, then feed and must molt twice to reach adulthood. The adult males are smaller than the females and have wings. They fly off in search of as many mates as they can find for the couple of days they have to live. The adult females instead secrete a brittle wax that covers the back to make a protective dome. This is the scale.

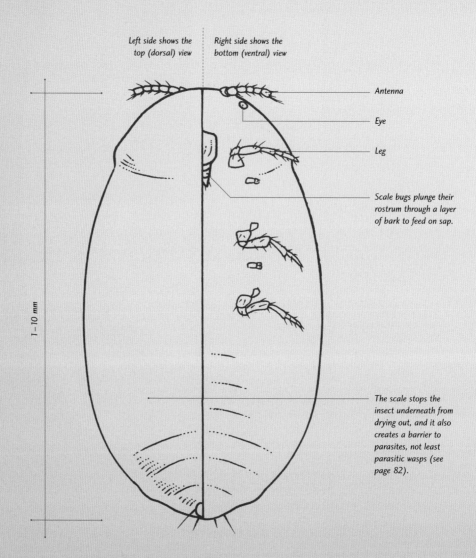

Left side shows the top (dorsal) view

Right side shows the bottom (ventral) view

Antenna

Eye

Leg

Scale bugs plunge their rostrum through a layer of bark to feed on sap.

1–10 mm

The scale stops the insect underneath from drying out, and it also creates a barrier to parasites, not least parasitic wasps (see page 82).

FIG. 1 WAX DOME

WAX AND FOAM

FEEDING FROM SAP

Bugs typically feed on plant sap, which is mostly water with a few sugars and other nutrients mixed in. As such, plant bugs, which also include aphids, whitefly, froghoppers, and cicadas, need to drink as much as they can as fast as they can. Froghopper nymphs will attack fleshy green plant stems.

Scale bugs target woody trees and have to plunge their rostrum, or needlelike mouthparts, through a layer of bark. Once the insect has established a feeding position, it would prefer to stay where it is, and so that is what scale insects—the females, at least—do.

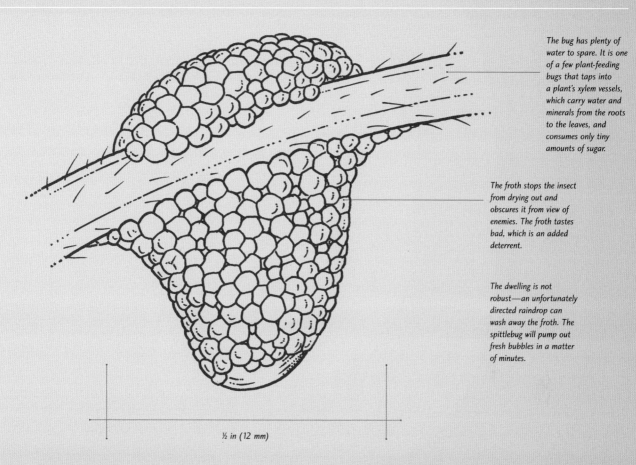

The bug has plenty of water to spare. It is one of a few plant-feeding bugs that taps into a plant's xylem vessels, which carry water and minerals from the roots to the leaves, and consumes only tiny amounts of sugar.

The froth stops the insect from drying out and obscures it from view of enemies. The froth tastes bad, which is an added deterrent.

The dwelling is not robust—an unfortunately directed raindrop can wash away the froth. The spittlebug will pump out fresh bubbles in a matter of minutes.

½ in (12 mm)

FIG. 2 FROGHOPPER BUBBLES

Froghoppers are plant-feeding bugs, so-called as the green nymphs make impressive leaps and have a squat appearance. The adults mate in late summer and lay eggs on the stems of mostly herbaceous plants. The eggs lie dormant over winter and hatch in spring. Froth then appears on the stems—not, as folklore suggests, formed from the spit of cuckoos, but from bubbles created when the nymphs mix their urine with gelatinous secretions from anal glands, then take in air from their spiracles (breathing pores). Inside the white mass is a froghopper nymph, or spittlebug.

FIG. 2 FOAM HOUSE

MATERIALS AND FEATURES
Ambrosia Beetle Gallery

Woodworms are not worms but the grubs of various kinds of beetle. They may attack dead wood, including furniture and timbers in a home, or the wood and bark of a living tree. The beetles that attack living trees are mostly weevils. These generally small beetles, measuring barely 1 millimeter long, have elongated snouts that contain highly efficient chewing mouthparts.

Bark beetles (Curculionidae) are weevils that dig mostly into the softer tissues under the hard outer bark. The family also includes ambrosia beetles, which create a fungus garden inside the tree to provide food. One Australian species, *Austroplatypus incompertus*, has evolved a bizarre social structure that closely resembles the cooperative communities of termites and ants.

This same species is the only beetle known to adopt a eusocial lifestyle, where a single female breeds, while many close relatives devote their lives to ensuring the survival of their kin—a way of life among termites, ants, and some bees, wasps, and even rats. A breeding female can live for ten years.

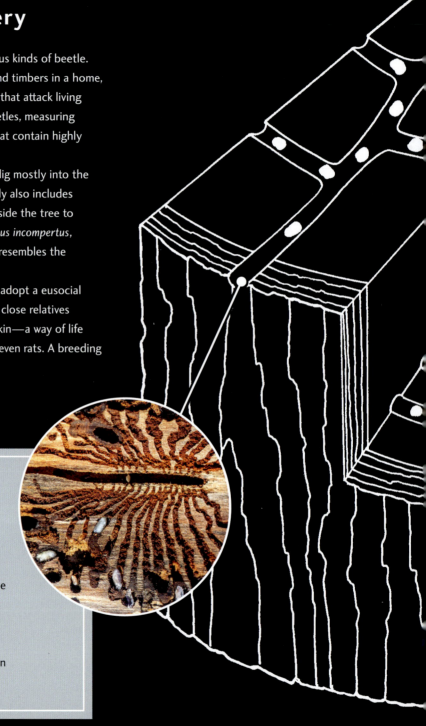

Digging chambers
An ambrosia beetle gallery is created by a founding adult boring its way into the heartwood of a eucalyptus tree. Generally this is a female, who is joined soon after by a male mate. However, in species where a male has several mates inside the tree, he will be the first to start digging. The parents create a central gallery then, after hatching, the larvae dig sideways, creating a distinct pattern of chambers.

Ambrosia fungus

After mating, the founding female spends several months boring a horizontal tunnel into the heartwood of a tree, spreading ambrosia fungus, carried in pouches called mycangia, as she goes. This fungus becomes established as she lays eggs at the far end of the tunnel. The grubs that hatch eat the fungus that grows around them. The mother must stay inside to control the fungus' spread, which, if left unchecked, could inundate the gallery.

Workers

The larvae can take up to four years to mature. The males leave the gallery first, flying away to find a mate in the outside world. Some of the females follow, some stay. Like their mother, they soon lose the lower segments of their legs, making it easier to move in the galleries but precluding them from a life outside. These worker daughters will not breed but maintain the fungus and care for the next round of their mother's eggs and grubs. Colonies have been found with dozens of workers caring for up to a hundred larvae.

BEETLES AND BUGS

Carrion Beetles Dig a Grave

Carrion beetles (Silphidae) eat the remains of dead animals. The brightly colored insects, mostly found in cooler parts of the world, are also known as sextons, undertakers, and burying beetles. While some of the two hundred species simply lay their eggs close to a large carcass so their grubs can feast on the flesh, others target smaller food sources, such as dead squirrels or birds, and dig graves for them. The parents then treat the carcass to make it easier for their larvae to eat. Habitat destruction and the over use of pesticides is putting carrion beetles under threat, in North America especially.

1 Gravediggers ↑
A male and female beetle find the remains of a recently dead mouse. This provides enough food for up to thirty grubs. It is a race against time to secure the food source, and the 1-inch- (3-cm)-long beetles begin to dig out the soils from underneath the carcass, so it steadily sinks below the surface covered over with heaps of soil on top.

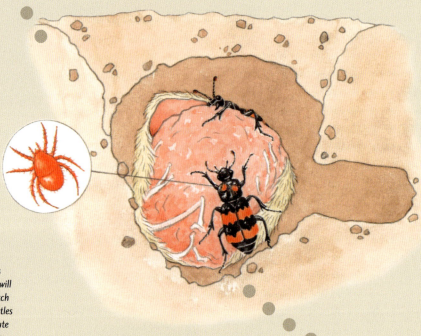

2 Skinning ↗
Now the carcass is underground it is at less risk of attracting flesh flies, which will lay their eggs on it. The maggots that hatch would be unwelcome competition. The beetles carry with them symbiotic mites. The minute predators now move to the carcass to prey on any fly eggs and maggots that have gotten through. Meanwhile, the beetles strip away the fur to expose clean flesh.

3 Preparing the food
The beetles coat the skin and flesh of the carcass with secretions from their mouth and anus. These contain preservative chemicals and a microbiome of bacteria that works to slow the decay of the meat and reduce its smell, which would attract more flies and larger scavenging animals.

CARRION BEETLES DIG A GRAVE   23

5 Feeding the young ↘
At first the grubs are not able to tackle the carrion diet alone. Instead, they feed on a nutrient-rich liquid regurgitated by the parents. The parents are themselves eating the carcass, all the while keeping it clean and well prepared. When the liquid meal is ready, the parent releases a chemical signal which makes the grubs adopt a begging posture so they can be fed. This mouth-to-mouth feeding ensures the symbiotic microbiome used to prepare the flesh is passed on.

6 Going it alone
The adult male is the first to leave the grave, generally around the time that the grubs are large enough to feed for themselves after a few days. The adult female stays for a few more days, guarding the grubs until they are ready to pupate at the age of about ten days. She leaves at that point.

7 Young adults emerge ↑
Carrion beetle grubs are slow to metamorphose in the soil. They take about four to six weeks to reconstitute themselves into the adult beetle form. They then dig themselves out of the grave and set off to find a mate. The entire process of burying the carcass to new adults appearing takes less than two months.

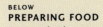

4 Producing eggs ↓
The beetle pair are now ready to mate. The female digs a short side tunnel out from the main chamber and lays her eggs there. The grubs hatch after about four days.

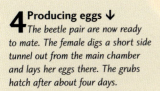

**BELOW
PREPARING FOOD**
A carrion beetle coats the skin of the carcass with a fluid that kills bacteria and prepares the flesh for consumption by the young.

BEETLES AND BUGS

Rolling a Ball of Dung

For eight thousand species of beetles in the superfamily Scarabaeoidea, dung is the only thing on the menu. These dung beetles generally target the dung of large herbivorous mammals, since around half of the nutritious material in the grass and leaves they eat is still there when expelled. Dung beetles eat dung themselves, but they also ball it up and provide a healthy scoop for their grubs to eat, either by tunneling directly under the dung ball and laying an egg inside it, or rolling it away and burying it in a more secluded spot. The two types of beetle—those that tunnel and those that burrow—look decidedly different.

1 Smell race ↑
Dung beetles are mostly nocturnal. When they are on the hunt, they look out for the freshest, softest dung, and sniff the air with their feathered antennae while patrolling on the wing. Once they get a whiff, the race is on to beat other beetles to the prize—and other insects that might contaminate it.

2 Cut and measure
Like their burrowing cousins, ball-rolling beetles have toothlike projections in their forelegs which they use to cut a chunk of dung from the pile. This is easier to do with fresh droppings, which have yet to harden up. The beetle periodically measures the ball it is creating by spreading its back legs. Around the same width is optimal; if the ball is any broader, it becomes too cumbersome to bury.

3 Rolling! ↑
The ball is around ten times the weight of the beetle. The insect turns backward, pushes the ball with its long back legs, and walks on its forelegs. If at any point it stops, another beetle may come and steal the ball. So, the beetle moves fast, traveling in a straight line as the quickest way to leave the area. The beetle maintains its path using the location of the sun or the orientation of the Milky Way.

ROLLING A BALL OF DUNG

5 Brood ball ↘
A mated pair of beetles work together to create a brood ball for their grub. The male does the rolling, while the female guards it from parasitic wasps that might lay an egg inside. Once the pair have wiped off fungus and removed any eggs laid by flies, the female coats the ball in her saliva and waste, lays one egg on it, and seals off the burrow. The male departs, while the female stands guard until the egg has hatched.

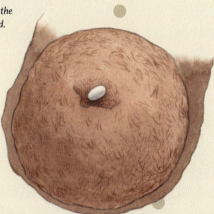

6 In the muck ↑
After hatching, the beetle grub burrows into the dung ball and eats its fill from the inside. It grows slowly, taking several months to reach a size ready for pupation and metamorphosis into an adult.

4 Burying ↓
Once the beetle reaches a patch of soft soil, it digs a burrow, frequently comparing the tunnel's width with that of the ball to ensure the dung fits inside. Once the ball is concealed, the beetle can eat it at its leisure, safe from attack.

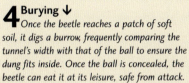

BELOW
DOR BEETLE
This dung beetle, a burrower, lives in meadows and farm pastures across Europe.

CASE STUDY
Twig Girdler

This species is a kind of longhorn beetle, so called because the insect's antennae are longer than the rest of its body. All longhorns feed on plant tissue, living and dead. The grubs mine away inside stems and trunks, causing a lot of damage. Twig girdlers are no exception: despite attacking living hardwood trees, the larvae eat dead wood. Their mothers construct a nursery filled with freshly killed wood that will be their home for months, if not years.

GIRDLING A TWIG
Adult twig girdlers appear in late summer and early fall. They are around ½ inch (1.5 cm) long, with antennae around ¾ inch (2 cm) long. The adults eat the tender shoots at the tips of twigs, living for around two months. In that time, they mate, and the females create unusual life-support capsules for their eggs. The female selects pencil-width twigs and cuts away a strip of the bark so there is a gap in the twig made only from wood. She then bores a little hole into the bark on the far side of the girdle before laying an egg inside. She lays one egg for every girdle—in total one female can lay upward of fifty eggs.

ROUND-HEAD BORERS
The twig's water and energy supply vessels are in the outer layers of wood beneath the bark, and so with that gone the girdled twigs die. They snap off at the dried-out sections, and the ground around an infested tree becomes littered with dead twigs. Inside, the beetle larva hatches after around three weeks and burrows into the wood interior. It leaves the outer casing of bark intact. The grub is known as a round-headed borer due to its obvious and hardened mouthparts at the head-end of an otherwise featureless tubular body. With winter looming, the borer lies dormant inside its twig-shaped capsule, only beginning to feed in earnest come the spring. It eats its way toward the girdled end, filling the mined cavity left behind with frass (waste) and wood fibers. In warmer climates, the larva pupates for fourteen days and bores its way out of the twig in the first year, emerging in late summer. In cooler regions, twig girdlers take two years to fully mature.

Classification

ORDER	Coleoptera
FAMILY	Cerambycidae
SUBFAMILY	Lamiinae
SPECIES	Oncideres cingulata
DISTRIBUTION	North and South America
HABITAT	Forest
NEST MATERIAL	Twigs
DIET	Dead wood

TWIG GIRDLER 27

ABOVE
UNDER ATTACK
This branch has clearly been attacked by longhorn beetles.

PEST PROBLEM

Twig girdlers target a wide number of forest trees, but they also attack orchard varieties, too, most notably pecan trees. An infestation of these beetles is a significant problem because the loss of twigs reduces space for fruits that year. In the following years, the loss of twigs unbalances the symmetry of an otherwise well-pruned orchard tree, which also reduces yields. To minimize attacks, farmers carefully collect any girdled twigs found in plantations and destroy them.

CASE STUDY
Tiger Beetle

Often showing off with metallic colors, or otherwise a uniform black, tiger beetles are the sports cars of the insect world. They are built to move fast and hold the land-speed record for any insect. One Australian species clocks in at 5½ mph (9 km/h), which is fast when the runner is only ½ inch (14 mm) long. The adult beetles use their speed to chase down prey, killing it with a slice from its scissorlike mandibles.

Classification

ORDER	Coleoptera
FAMILY	Carabidae
SUBFAMILY	Cicindelinae
DISTRIBUTION	Worldwide
HABITAT	Areas of sandy ground
NEST MATERIAL	Sand and soil
DIET	Ants and spiders

VERTICAL SHAFT
After mating, an adult female tiger beetle digs several short vertical burrows into soft, sandy soils. She then lays a single egg in each one. When the larva hatches, it extends its burrow, tunneling down to create a refuge from bad weather, and the larva widens and deepens its lair as it grows. Some eventually burrow 6½ feet (2 m).

ATTACK MODE
When conditions are right to catch a meal, the tiger beetle larva climbs to the entrance of its burrow. The opening is a good fit for the insect's head, which is held flush to the surface. The head forms a lid that seals the burrow, obscuring its presence. Behind its head is an armored section, called a promotum. The larva sits at the entrance for hours, held in place by a hook that pokes out of the top of its abdomen.

ESCAPE STRATEGIES
The tiger beetle larva's hideout is largely impregnable. If a big predator comes near, the insect disappears out of reach. The only persistent threat comes from parasitic wasps, with each species targeting a particular tiger beetle, aiming to lay eggs inside it. These wasps are small enough to follow the beetles into the den. The larva cannot run away, only wriggle. However, the eastern beach tiger beetle (*Habroscelimorpha dorsalis*) larva has evolved a way to escape: it springs out of its nest, curls its long body into a circle, and rolls away, relying on the ocean breeze to push it on its way, beyond the reach of the wasp.

TIGER BEETLE 29

LEFT
KILLER LARVAE
Tiger beetle larvae are active predators that construct a hideout for ambushing prey, using their own heads to seal off its secret entrance.

CASE STUDY
Periodical Cicada

Cicadas are among the largest and most familiar members of the bug order. Their clicking songs are some of the loudest sounds produced by any insect, and cicadas often sing together in their thousands, filling the air with a deafening rhythm. These are the calls of the adults, made after they emerge from a long stretch underground as nymphs. The wingless nymphs live a sluggish existence in soil, sucking on sap tapped from tree roots.

Classification

ORDER	Hemiptera
FAMILY	Cicadidae
SUBFAMILY	Cicadettinae
SPECIES	*Magicicada septendecim*
DISTRIBUTION	Eastern North America
HABITAT	Broadleaf forest
NEST MATERIAL	Mud
DIET	Plant sap

PRIME NUMBERS

Periodical cicada nymphs can typically be found 24 inches (60 cm) underground, drinking the xylem sap from tree roots. This is a poor food source and these bugs take many years to reach maturity. When that time comes, however, hundreds of thousands climb out of the ground on the same night and complete a final molt to become fully fledged adults. A predator that can synchronize its life cycle to that of these slow-growing bugs would be devastating, and so this genus of cicadas employ the prime numbers thirteen and seventeen to make it difficult for predators to track their development. So, some species emerge after thirteen years underground, while others always wait seventeen years. There is no shorter life cycle, other than a general yearly one, that a predator could follow to ensure they are always around at the same time as the vast swarms of cicadas. And no predator has taken on a thirteen- or seventeen-year cycle for this purpose—yet.

MUD TURRETS

Mature nymphs emerge in spring, whenever the temperature aboveground rises to around 64°F (18°C). In preparation, the nymph digs a tunnel to the surface and waits. Often the tunnel is extended by a turret made from heaps of mud. The turrets can be 4 inches (10 cm) tall, and to begin with they have a sealed top, and so are also called cicada "huts" or "cones." The structures are most commonly seen in damp soils where rainfall is high. Their purpose is to stop the nymph's escape tunnel being flooded and to provide an air snorkel if the upper layers of soil become waterlogged.

PERIODICAL CICADA   31

LEFT
EMERGENCE
When the ideal external temperature is reached, the cicada waits until dusk to break open the top of the tube. As a result, the mud structures are normally seen as turrets, with open tops. The cicada that made them has long gone, hastening up into the nearest tree to begin its adult life. After years underground, the cicada has only a few, very noisy, weeks of life left.

ABOVE
RECENT ARRIVAL
Periodical cicadas spend years down in the soil, keeping predators guessing as to when they will show themselves at the surface. Before their arrival, they build a turret of mud. The turrets pictured here have broken tips, showing that the insects have recently emerged from the tops.

CASE STUDY
Blister Beetle

The adult form of this small flower beetle is unremarkable. The 10-mm beetle has a metallic sheen to its dark body. It eats little, surviving on pollen and nectar from scant desert plants, as it searches for mates. That phase of life is without much drama compared to what happens next. The grubs grow inside a bee nest, feasting on the pollen and nectar provided by their hosts but also the eggs and larvae. How they get there is one of nature's most amazing stories.

BEE MIMIC
A blister beetle develops through two main larval forms, first a triungulin, a highly active phase, and then this transitions into a more typical grub. The eggs hatch and tiny, wriggling triungulins emerge en masse. They climb up a nearby plant stem and use their own bodies to build a structure that looks remarkably like a bee perching on the stem; this is where mimicry and purpose-built structures meet. The mass of triungulins then emits the chemical scent of a female digger bee, and eventually (it can take a few days), a male digger bee comes to mate. However, his mating attempt results only in many of the triungulins clinging to his body as he flies away to find a more suitable partner. Assuming he is successful, the hitchhiking beetle larvae then jump ship to the female bee, which unlike a male, eventually builds a nest, gnawing living spaces in soft rocks. This nest is where the triungulins want to be. They disembark and through a series of molts transform into a grub form more commonly associated with beetles.

BURNING OILS
Blister beetles are named for a defensive tactic that sees the adults exude an oil from their leg joints when threatened. (An alternative common name for the family is "oil beetle" for this reason.) This active chemical in the oil is a fat-based substance called cantharidin. It is an irritant that can cause blisters or other irritations when it contacts the skin. These blistering abilities have seen the beetle oil used for corroding away warts and similar skin blemishes.

Classification

ORDER	Coleoptera
SUPERFAMILY	Tenebrionoidea
FAMILY	Meloidae
SPECIES	*Meloe franciscanus*
DISTRIBUTION	Southwestern United States
HABITAT	Sandy areas
NEST MATERIAL	Living bodies
DIET	Pollen and nectar, bee larvae

BLISTER BEETLE   33

ABOVE
HITCHING
A mass of triungulins on the back of a digger bee wait to be taken to the nest.

LEFT
DECEPTIVE POSE
After hatching, a mass of triungulins crowds to a high point and forms a mass that resembles—and smells like—a female digger bee.

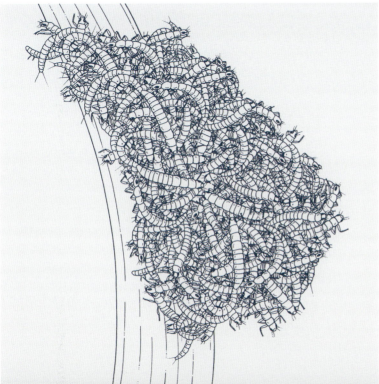

CASE STUDY
Common Earwig

This species of earwig originates in Europe and western Asia but was introduced across the world in the twentieth century and is now established in every continent (bar Antarctica). In common with other earwigs, this species gets a survival boost from the high level of care the mother gives her young. This plays out in a hidden subterranean chamber constructed by both parents.

Classification

ORDER	Dermaptera
FAMILY	Forficulidae
SPECIES	*Forficula auricularia*
DISTRIBUTION	Worldwide
HABITAT	Temperate and tropical areas
NEST MATERIAL	Soil and leaf debris
DIET	Insects and plant material

WINTER LAIR
Adult earwigs are loners, and mates pair up for a short while in late summer. They work together to dig a shallow, tube-shaped burrow in soft soils, but then the male is chased away by the female. Once alone, she lays about fifty eggs inside. The eggs stay in the chamber through winter, away from the frosts. In that time the mother stays with them, in a largely dormant state. She periodically turns the eggs, cleaning away any fungus that has appeared and, as the conditions change, she moves them around the chamber, piling them up wherever the temperature and humidity is optimal.

FEEDING TIME
In spring, the mother spreads out the eggs as they approach time for hatching, staying to guard the tiny nymphs that emerge. They are wingless, but already share the distinctive earwig shape with pincerlike cerci at the rear.

A MISLEADING NAME
The earwig's most obvious features are its long, pincerlike cerci at the back of its abdomen. These are longer in the males, and used as a signal of reproductive vigor, but both sexes use their cerci as defensive weapons when needed. The name earwig has been conflated with the cerci in some cultures to promote the myth that earwigs climb into ears and burrow into the brain. This is of course nonsense. Earwig does mean "ear insect" in Old English, but it earns that definition from the shape of its small wings, which resemble human ears.

COMMON EARWIG   35

ABOVE
NESTING
The mother feeds her babies by regurgitating her own partially digested food and later by delivering food items to the den. She stays with them for about fourteen days, until they have undergone their first molt. From then on, the babies are on their own.

CASE STUDY
Mole Cricket

These orthopterans spend their days tunneling. From egg, through nymph, to adult, the mole cricket is mostly hidden away within an extensive tunnel network in the soil. The prairie mole cricket from the tallgrass prairies of the United States is one of the biggest—and noisiest—examples. During the mating season female mole crickets take to the air in search of a partner, and the males construct horn-shaped burrows that broadcast their love songs.

Classification

ORDER	Orthoptera
SUPERFAMILY	Gryllotalpoidea
FAMILY	Gryllotalpidae
DISTRIBUTION	Worldwide
HABITAT	Moist ground
NEST MATERIAL	Soil
DIET	Grass roots

SINGING BURROW

When the breeding season arrives on the prairie in spring, male mole crickets each dig themselves a "singing burrow." These are acoustic horns carved into the soil. The cricket stands, head facing down, with the front part of the body in the "bulb," a resonating chamber that is around ¾–1¼ inches (2–3 cm) underground. The bulb connects to the surface via the horn through a tight opening (or throat) that is about ⅔ inches (1.5 cm) across. The cricket's wings and rear end pass through the throat to sit in the lower portion of the horn. The horn's length is about 3½ inches (9 cm); the width is 1 inch (2.5 cm). These dimensions are best for amplifying the cricket's 2 kHz chirpings, which he makes by scraping his forewings together. The calls can be heard clearly from ¼ mile (400 m) away.

CRICKET ORCHESTRA

The prairie mole cricket is a lekking species. This means that rather than avoiding the competition, rival males deliberately gather together. They build their singing burrows in an arena, listening out for the sounds of activity vibrating through the soil. This species breeds more readily when the grass is still short. The burrows have to be more widely spaced in locations where the grasses are higher, and the horn shapes are also directed more upward to reach the females higher overhead. All females in the air will be drawn to the arena due to the mass calling.

MOLE CRICKET 37

LEFT
EGG STORE
The female uses her shovel-shaped forelegs to dig a deep pit for her eggs.

BELOW
LOUDHAILER
The trumpet-shaped opening of the mole cricket's burrow helps to amplify the mating calls.

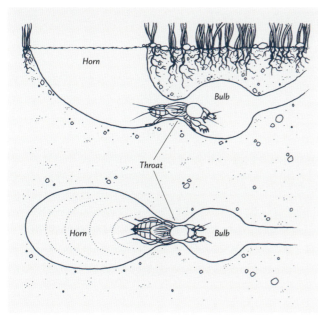

SPADE LEGS

Mated females dig a deep burrow in the soil, down to around 1 foot (30 cm) or more. They will lay around fifty eggs inside, and then dig a burrow next door to stand guard over them until they hatch after a few weeks. The nymphs, like the adults, eat the grass roots that permeate the soil here year-round, even if the plants have died back aboveground. To reach the roots, the crickets need only to dig. The forelegs of the mole cricket, in nymph and adult alike, are well adapted for digging. The lower segment is flattened into a shovel shape, with chunky serrations that cut into the soil. Each spadeful is then pushed back as the cricket moves forward. Mole crickets are very fast diggers. They can even evade an imminent attack on the surface by digging an emergency escape tunnel.

CASE STUDY
Ice Crawler

This small and drab wingless insect is easy to overlook. However, it does something quite amazing for a so-called cold-blooded animal: it builds tunnels inside snow and ice. "Cold-blooded" refers to the way the insect has few, if any, internal mechanisms to control its body temperature. Instead, it relies on external sources of heating or cooling, and essentially is the same temperature as its surroundings.

ON THE SNOW LINE
The ice crawler is around 1 1/4 inches (3 cm) long, but that measure is longer if the spiked cerci that protrude from its abdomen are included. Found along the edge of the snow on high rocky slopes, ice crawlers have only small eyes—although these are big compared to those of most other species in the same family, many of which are completely blind. Nevertheless, this nocturnal insect relies mostly on its long, flexible antennae, which are covered in sensitive hairlike projections, to find food. It is mostly a scavenger, patrolling the snow and ice for the remains of other arthropods.

RESTING PLACE
During the day, the ice crawler comes off the snow field and retreats under rocks. If the temperature rises too high, the crawler digs down into colder soil to stay cool. It is most active when temperatures are hovering around freezing. In winter, the surface temperature in the mountains will drop well below that, and so the ice crawler tunnels under the deep snow and continues to feed on what it can find. It is still very cold down there, but the snow layer acts as insulation, stopping the temperatures dropping to dangerous levels. The ice crawlers' sub-zero home is very simple. The insects are slow movers thanks to the cold, and little is known about their life under the snow. They use their long body to wriggle into narrow spaces.

Classification

ORDER	Grylloblattodea
FAMILY	Grylloblattidae
GENUS	*Grylloblatta*
DISTRIBUTION	USA and Canada
HABITAT	High mountains
NEST MATERIAL	Rock, snow, and ice
DIET	Snow flies and mites

FREEZING MYSTERY

No one really knows how ice crawlers can survive in such cold conditions. Other cryophilic animals are able to withstand freezing conditions by having high concentrations of antifreeze chemicals like sorbitol and glycerol in their body tissues. These stop large ice crystals from forming in the body which would damage vital tissue. However, ice crawlers do not have these chemicals. They will die from the effects of ice when the temperatures are low enough. They grow more sluggish below 23°F (-5°C), become inactive at 21°F (−6°C), and are dead at 16°F (−9°C).

ABOVE
COLD-LOVER
An ice crawler out and about at night on the ice. The ice crawler lives up to the term cold-blooded by being a cryophile (cold-lover). It builds its home in ice and will die if it gets warmer than 68°F (20°C), so even the warmth of your hand can be deadly.

CHAPTER TWO

Web Spinners and Silk Weavers

Silk is a building material like no other. Each strand is formed from a droplet of protein that solidifies on contact with the air. The protein takes its final form when put in tension, becoming a kind of crystal with great strength. That transition is not instant, and in the hands—or rather appendages—of a skilled animal architect, the silk is pulled into strands that are used to weave, wrap, and glue, creating a wealth of structures.

Spiders are the animals most celebrated for their silk workings. Most obviously we think of their intricate webs, set up to trap prey, but spiders also build sacs for eggs, funnels as nests, and they shroud prey in silk, saving them for eating later. However, spiders are not insects, of course. They are not even that closely related, with the two groups diverging somewhere on the seabed of the Cambrian period more than 500 million years ago. Insects are terrestrial descendants of shrimplike crustaceans that climbed out the water about 100 million years later. None of the extant marine relatives of insects or spiders use silk, and the material we see today does not lend itself to use in saltwater, although insects and spiders do make it in freshwater habitats. So it is very likely that insects and spiders evolved their processes of silk production independently of one another. Indeed, there is overwhelming evidence to back this up. Spiders are able to tailor the material properties of their silk to meet different needs beyond mere construction, whereas an insect makes do with a single kind of silk. More significantly, spiders exude their silks from spinnerets, which are modified appendages on the abdomen, connected to internal silk glands. By contrast, insects that make silk use glands located in a range of different body parts, thus indicating that insect silk has evolved independently several times.

SILK GLANDS

Many insect orders have the ability to create silk. The silk builders most familar to us are are the moths and butterflies (Lepidoptera), many of which spin a silken cocoon as they prepare to pupate and metamorphose from caterpillar to winged adult. The fabric we call silk is harvested from the cocoons of moth caterpillars, or silkworms. The finest and most highly prized silken garments, which are both strong and lightweight, are made from the fibers of the silk moth (*Bombyx mori*). However, high-quality silks are also made via the industrial exploitation of other moth species.

Lepidopterans make their silk using modified salivary glands linked to the labium, the structure that forms the base of an insect's mouth. Most insects that make silk use these labial glands. Bark lice (Pscoptera), which weave large

OPPOSITE
SILK MAKER
A silk moth emerges from a pupal case cocooned in strands of woven silk.

communal webbings, use these glands, as do flies (Diptera), caddisflies (Trichoptera), and wasps, ants, and bees (Hymenoptera).

However, beetles (Coleoptera) and lacewings (Neuroptera) make silk with glands that have evolved in the Malpighian tubules. These are the equivalent of the insect's kidneys, with the role of excreting waste materials from the body. A few orders, such as the mantids (Mantodea), use glands in the genitals to create silk for weaving egg sacs. Thrips (Thysanoptera), or thunder flies, produce silk with their anal glands. Meanwhile, web spinners (Embiodea) have developed silk-producing glands on their feet.

ABOVE
SILK GLAND
This image shows a close-up view of the silk glands of a saturniid moth (also called giant silk moths).

OPPOSITE
WEB SPINNER
A female web spinner maintains a silken shelter for her eggs.

WHAT IS SILK?

Insect silk is made from a protein called fibroin. This is a polymer of smaller amino acids chained together. These chains form into sheet-like structures (while other proteins take on more blobby globular forms). The silk gland produces a starting gel of fibroin and water. This receives a coating from other proteins as it moves through the gland to the spinneret. These work in pairs pushing the gel through tiny holes so it forms threads. On contact with the air, the proteins begin to crystallize, and every strand of silk is created by spinning a pair of threads together to boost strength. Silk is a costly product to make, and so generally it is larval insects, which have (hopefully) a ready access to nutrients and more time to exploit them, that make it in the largest quantities. Once these insects have metamorphosed into adults, silk production becomes less common. However, several hemimetabolous insects use silk throughout their lives, when young and old.

USES FOR SILK

The way lepidopterans use silk is familiar enough. The silk strands are used to bind a hook-shaped structure, called the cremaster, on the tip of the caterpillar's abdomen to a branch or other secure structure. Some attach the cremaster with a sticky silken pad; others tie themselves on. This holds the insect in place as it enters the immobile pupal stage. The caterpillar then generally wraps and weaves a cocoon of silk strands around itself before pupating. (Many lepidopterans don't use the cremaster but cocoon themselves under the soil.)

The skin of the caterpillar thickens, forming a pupal case, and inside the body is transformed from a larva, intent only on eating, to a winged adult with the task of finding a mate. When that stage is completed, the dried cocoon splits, and the moth or butterfly emerges. It takes a few minutes to pump body fluids into its wings and other new body parts, harden up, and then flutter away.

This chapter looks at how some lepidopterans and other unrelated insects go further than building a pupal cocoon or protective case for their eggs. The web spinners, as readers may already have surmised, construct extensive silken tunnel networks, creating a communal fortress where many insects can feed undetected. Several kinds of barklice, the wilder cousins of booklice, construct much larger webbed encampments in trees. And finally, thrips are frequent users of silk, and some species set up home in *Acacia* trees, constructing safe breeding chambers by lashing together their needlelike leaves. Silk is a wonder material. Let's see what insects can build with it.

Silk from Silkworms

The silk of silken garments is an insect building material, specifically the creation of the caterpillars of *Bombyx mori*, the silk moth, which uses it to construct a cocoon around the pupa. The pupal case is built from a single, ultrafine thread of silk which can be almost 1 mile (1.6 km) long. The silken cocoon may be simple in structure but the woven thread creates a protective layer around the pupa. The silk from this moth has been produced in China for around five thousand years, and its cultivation was a closely guarded secret for centuries. So highly prized was this luxury fabric that the main East–West trade routes of the ancient world were dubbed the Silk Road and supported entire civilizations.

3 Growing up →
The silkworm molts five times and increases in length eightfold by the time it reaches the final instar, or larval stage. The caterpillar will be between 21 and 25 days old by this time, and about 2¾ inches (7 cm) long.

1 Egg production ↓
The female moths are heavier than the males. Only the healthiest adults are allowed to breed. After mating, a female takes around twenty-four hours to be ready to lay eggs. If kept in optimum conditions, one moth will lay five hundred eggs. The eggs have a sticky coating, so they cling to the upper surface of mulberry leaves.

2 Silkworms ↑
The near-spherical eggs hatch after about ten days. The caterpillars eat continuously and will die if there is not enough food. Traditionally, silk workers fed the silkworms on fresh mulberry leaves. Today, the leaves are generally substituted with a prepared food source.

SILK FROM SILKWORMS

4 Spinning a cocoon →
The silkworm stops eating once it reaches full size. Its salivary glands are now preparing to produce silk. The caterpillar seeks out a secluded spot and starts to weave a silken cocoon around itself by swiveling the head. This can be achieved with a single thread of silk 1,600–5,000 feet (500–1,500 m) long. This weaving process takes about five days.

5 Removing the pupa
If left to complete its life cycle, the pupa will keep one end of the cocoon moist and soft, while the rest hardens. After a month or so, the adult bites through the soft end. However, in industrial silk production, the adult moth is not needed, and the older cocoons yield fewer strands. So, once the cocoon is complete, it is dropped into hot water. The heat kills the insect and softens the silk filaments so they are easier to separate.

RIGHT
SILK STRANDS
A single thread of silk used to make a cocoon is only 15 micrometres (millionths of a meter) wide. The strand is thickened considerably, however, by coatings of waterproofing chemicals.

WEB SPINNERS AND SILK WEAVERS

BLUEPRINTS
Webbing and Cases

The clothes moth and carpet moth are two near-ubiquitous pest species that plague homes the world over, targeting carpets, clothes, and other soft furnishings. They are members of the Tineidae family, collectively known as the fungus moths, and in the wild they are detritivores that scavenge for scraps of waste along with the fungus that grows on them. Some target the keratin protein in natural animal fibers, such as hair and feathers, and can be found in the nests of hibernating mammals and the roosting areas of bats and birds. Unfortunately, modern homes, filled with wools and other natural fibers, are a cornucopia for these moths.

FIG. 1
WEBBING OF DECEPTION
The clothes moth is also called the webbing clothes moth because the caterpillars, which are nearly too small to see, spin mats of silk to screen their presence.

They stay hidden underneath until night falls and then wriggle out to chomp their way through woolen clothing, carpets, or soft furnishings.

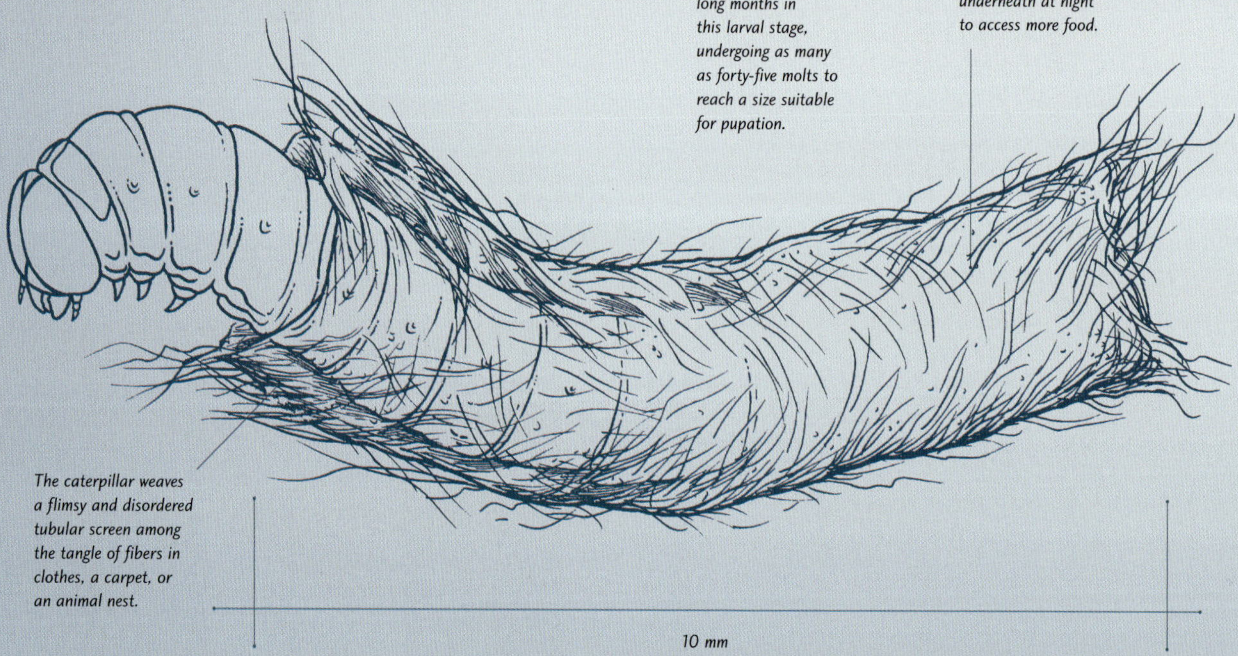

The moths spend long months in this larval stage, undergoing as many as forty-five molts to reach a size suitable for pupation.

The messy silk screens cover daytime activity, and the caterpillar may wriggle out from underneath at night to access more food.

The caterpillar weaves a flimsy and disordered tubular screen among the tangle of fibers in clothes, a carpet, or an animal nest.

10 mm

FIG. 1 CLOTHES MOTH LARVA

THE CLOTHES MOTH

The golden-brown adults of the common clothes moth (*Tineola bissellietla*) are around 5 mm long and are easier to spot than the larvae (see below)—but the damage is already done by then; it is not the adults that feed on clothes but the larvae. The females barely move, perching in wait of the males, which scuttle around looking for them, making the occasional short, fluttering flight. Pheromone traps can snare the males but have no effect on the female.

At night, the caterpillar pokes its head out of the case to start to feed. It hauls its home behind it, ready to withdraw for its own protection if needed.

FIG. 2
MAKING A CASE

The case-bearing moth (*Tinea pellionella*) is also found worldwide. They look more silver-gray than the common clothes moth and are generally larger. Although they may be found in houses, these moths are less damaging, and it is less likely to be the species destroying your belongings. However, case-bearing moths are more advanced builders. The larvae collect the fibers around them and weave them with silk into a tubular case.

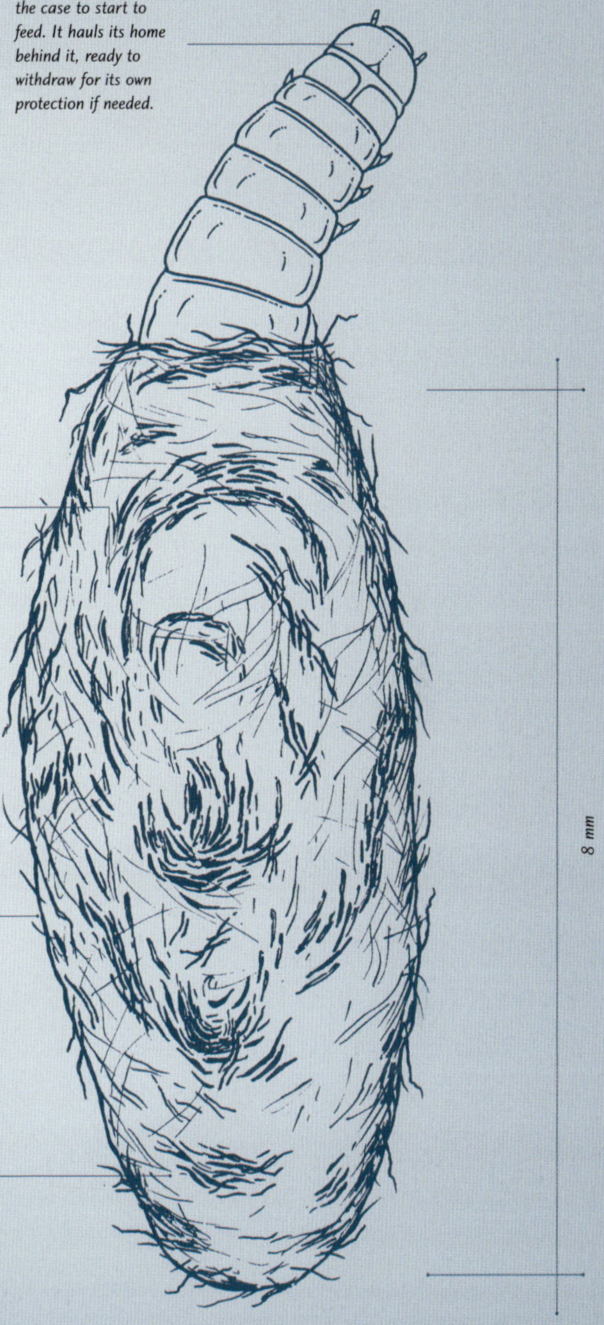

By day, the larva lurks inside the case, which is nearly impossible to see because it is made from the same material by which it is surrounded.

The tube is open at both ends and is wide enough for the caterpillar to turn around inside and eat from both ends.

The animal enlarges its case as it grows, and when the caterpillar is big enough to pupate, it seals up the case and transforms inside.

The larvae are undeterred by dyed fabrics and so the case can be a brightly colored affair.

8 mm

FIG. 2 CASE-BEARING CATERPILLAR

MATERIALS AND FEATURES
Bagworms

The bagworms (or bagmoths) of the family Psychidae are cousins of the fungus moth. They are plant eaters in the main, and so are not general pests like their cousins. The family is named after the way the caterpillars build themselves a protective case as soon as they hatch. By using local building materials, such as twigs, seeds, and lichen, these cases are highly effective camouflage.

The flying adults of this species are exclusively male. A male caterpillar, once mature, pupates and metamorphoses into a small, fluffy, dark-winged adult. They break out of the bag and fly away, sniffing out the pheromones of females nearby. However, the adult female bagworm remains flightless and is confined in her case for much, if not all, of her life. Following pupation, she has transformed but retained a near-larval form. As well as functioning sex organs, she has six true legs, but they are short, and her antennae are mere stumps. The adult female stays inside the case, luring mates with her chemical messages. In some species, the case provides safe passage for the female's eggs—providing a protective barrier as it passes its way through the guts of a predator.

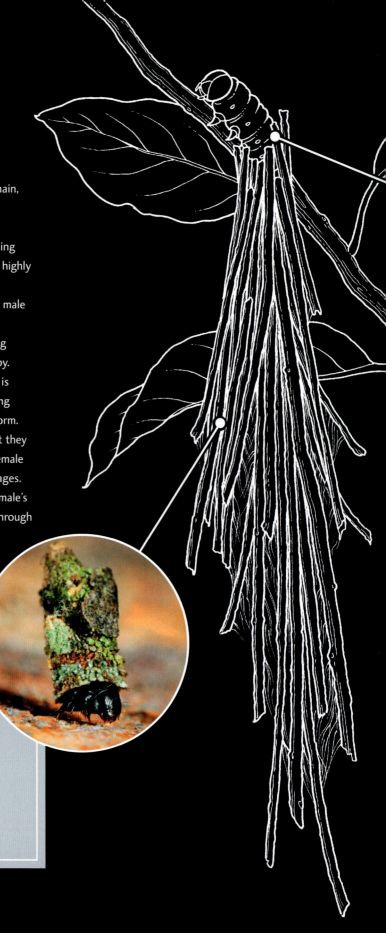

Decorative bag

The term "bag" for a bagworm's construction does not really do it justice. The edifice has a roughly tubular shape, but the use of natural building materials lends it a decorative look. This helps to break up a distinctive constructed shape so that the bag blends into the background.

Multifunctional case

On the inside the bag is lined with silk, but the outer surface is constructed with whatever comes to hand nearby. Often the case is made from twigs and stalks, lashed together, or scraps of lichen, seed capsules, or fluffy features and wool. The caterpillar hauls the case around with it and extends the bag as it grows. The head end of the case has the larger opening, big enough for the caterpillar to stretch out and graze. This opening can be drawn closed if the individual is attacked. The other end has a smaller hole for the removal of waste.

Attracting attention

After laying eggs, the females of many species simply die in the case. In other species, such as the dusky sweep (*Acanthopsyche atra*), the female shuffles partially out of the case and wriggles and writhes. She appears to be mimicking the movements of a chubby beetle grub or juicy maggot in order to attract attention from a predatory bird. She is likely to be eaten—along with her case and all the eggs. Many of her eggs survive the journey through the digestive tract of the bird and pass out in the droppings. This is likely to happen far from their starting place, so the next generation of moths can spread to new areas.

CASE STUDY
Rose Leafroller

This species is typical of the Tortricine, a subfamily of the Tortricidae, in that its members are small drab moths that mostly tunnel around inside their plant food sources, such as roots and stems. However, the Tortricinae are leafrollers, so named because the larvae target fresh leaves and buds and use silken strands to fold over neighboring leaves to create a secluded space for feeding.

GARDEN PEST
The rose leafroller is endemic to the temperate forests of Europe and Asia, and it is a common feature in cultivated gardens and orchards. Here it can be a pest of apples, raspberries, and garden roses, all of which belong to the same plant family Rosaceae. In the nineteenth century, the rose leafroller was introduced to North America, and it is particularly abundant today in the Pacific Northwest region and New England. Here it targets many Rosaceae plants, wild and cultivated, as well as attacking hazelnut plantations and privet hedges.

NEST BUILDING
The eggs are laid in early fall, clustered in mats of around fifty to one hundred on tree bark. The mats are protected through the winter from frost and desiccation by a gel coating. The larvae hatch in spring and they build a nest. The nest can be a leaf that is rolled over with one edge bound with silk to the opposite edge. The larva then feeds inside, stripping away the inner surface of the leaf. Other nests are created by tying neighboring leaves together to make a more irregularly shaped shelter. A giveaway sign that a caterpillar is inside a leafroller larva is the way it withdraws backward into its shelter if disturbed. If the threat persists, the larva evades the attacker by dropping out the bottom of the nest and dangling on a thread.

Classification

ORDER	Lepidoptera
FAMILY	Tortricidae
SUBFAMILY	Tortricinae
SPECIES	*Archips rosana*
DISTRIBUTION	Europe, Asia, and introduced to North America
HABITAT	Woodland
NEST MATERIAL	Leaves
DIET	Shoots and buds

ROSE LEAFROLLER 51

ABOVE
AT HOME
A rose leafroller caterpillar maintains the silk anchors that keep the leaf rolled up.

LIFE CYCLE

The moth produces a single generation a year. By the time it is big enough to pupate in early summer, the leafroller has enlarged its nest. It will have eaten away much of the earlier leaf walls and so other leaves have been pulled in to provide cover. It will also build around buds of new leaves and flowers so it can feast on these tender foods in peace. The caterpillar pupates inside its leaf nest, securing itself in place by silk webs. It takes around a month for the adult moth to appear.

CASE STUDY
Eastern Tent Caterpillar

This American species is a highly gregarious lepidopteran. Once the caterpillars have hatched they spend their larval stages in a tight-knit community. They build a communal nest—a tent—from webs of silk and all contribute to its construction and upkeep. Eggs are laid on the leaves of trees in spring and develop into tiny larvae within weeks but do not hatch. The larvae lie dormant over the winter and emerge en masse—250 or more—the following year.

SHARED TENT

The tents built by this species are among the largest of any tent caterpillar (a subgroup of lappet moths). Together the young caterpillars build a shared shelter high up in the tree, where several small branches fork out from a larger one. The main opening is at the top and the broadest wall faces toward the rising sun. (Experiments show that the caterpillars' construction behavior is highly dependent on the direction of light. If a strong light source is placed under the tree, then the tent is built upside down.) The residents add a new layer of silk to the nest each day. As this fresh layer dries, its strand contracts so it separates from the layer beneath. This creates a series of silk walls inside the nest. The caterpillars rest in the gaps in between.

TEMPERATURE CONTROL

As well as being a refuge for the caterpillars, the tent is also used to collect heat energy. As a result, the larvae inside are able to warm up and be active earlier in the year than is typical of other caterpillars. Tent caterpillars spend long hours basking in the tent. Their bodies are dark so as to absorb as much heat as possible. The temperature is slightly different between layers, and so the caterpillars switch location to moderate their body conditions. If the whole nest gets too hot, the caterpillars vacate it and hang on the outside of the shaded side so they can shed unwanted heat.

Classification

ORDER	Lepidoptera
FAMILY	Lasiocampidae
SUBFAMILY	Lasiocampinae
SPECIES	*Malacosoma americanum*
DISTRIBUTION	Eastern North America
HABITAT	Woodland
NEST MATERIAL	Silk
DIET	The leaves of cherry trees and other members of the rose family

EASTERN TENT CATERPILLAR   53

ABOVE
CLUSTERING
Caterpillars crowd together in the tent to share body heat.

WORKING TOGETHER

The caterpillars do everything together, including feeding. They have three meals a day, from dawn to dusk. When hungry the caterpillars leave the nest, following scent trails left by others in the days before that lead to food sources. The hungriest caterpillars are more likely to diverge from these foraging routes to find fresh foods. Once well fed, the caterpillars return and add more silk to the tent, thus ensuring that it grows in size along with them. If one caterpillar feels under attack from a predator, and especially from a parasitoid, they start to thrash their abdomens. The motion startles predators and makes it harder for them to focus on a single target prey. The behavior spreads, and soon the whole tent-load writhes and thrashes until the threat has passed.

CASE STUDY
Fall Webworm

As caterpillars, this species forms hundreds-strong colonies that live in communal nests constructed in the treetops from silken webs. With webs that can grow to around 3 feet (1 m) across, the webworms are highly destructive residents of trees and are considered a pest in many places.

Classification

ORDER	Lepidoptera
FAMILY	Erebidae
SUBFAMILY	Arctiinae
SPECIES	*Hyphantria cunea*
DISTRIBUTION	North America, introduced to Europe and Asia
HABITAT	Forest
NEST MATERIAL	Silk
DIET	Leaves

COMMUNAL WEB

The webs of the fall webworm start to appear in the middle of summer, after all the tent caterpillars have gone. The webs are only in use by the webworm larvae for around four to six weeks, but in that time, they spread out to cover several branches. The larvae leave the webs when it is time to pupate in the fall but leave behind a messy tangle of silk filled with fragments of leaves and the droppings and molted skins of several hundred caterpillars.

LIFE CYCLE

This moth species spends the winter as a dormant pupa buried in the soil or leaf litter. The adults emerge from their pupal cases at the end of spring. After mating, eggs are laid on the underside of leaves. The communal web takes shape around the feeding site, and so covers the leaves on a small branch at first and then spreads to neighboring branches as the larvae consume more and more leaves.

DETERRING DANCE

The web is constructed around the caterpillars' feeding site, and they will move across its exterior to find more food sources for expansion. The web keeps those inside warm and offers protection from predators that become tangled in its strands. However, if the web is threatened in some way, the webworms begin a jerking dance, rapidly twitching their heads back and forth in near unison. It is thought this mass movement startles predators into thinking that the web is in fact some kind of single, large organism.

LEFT
TEAM WORK
Several hundred eggs hatch together, and the larvae begin to feed en masse. The fall webworm is an unfussy eater and will target all kinds of deciduous trees.

CASE STUDY
Web Spinner

Web spinners begin spinning from their earliest days, and unlike many silk-building insects, they carry on building with silk for their whole lives. Mostly out of sight under rocks or among leaf litter, web spinners construct a communal home of crisscrossing tubular tunnels known as a gallery. The gallery is full of nymphs and their wingless mothers. Adult males have wings which they use to disperse and find mates.

FAMILY HOME
The web-spinner gallery is a family home, although it may contain several adult females and their young. Members of each family tolerate the presence of others, but do not cooperate or coordinate. Each adult cares for her own young, which means guarding them as eggs and nymphs. The adults and nymphs build the gallery to suit their own needs. Several tunnels radiate from a central chamber.

FOOT GLAND
The silk used in making the galleries is produced with a unique silk gland located inside a bulging segment of the web spinner's forelegs. The silk gel is exuded through hundreds of hollow hairlike setae under its two front feet. The setae function as nozzles and produce fine filaments of silk that form webs wherever the web spinner touches. Fresh silk is lavender in color, but it fades to white as it cures.

ADAPTED WINGS
Female web spinners remain inside these tunnels, while adult males grow wings and fly away, seeking another with female mates inside. The web spinner's wings have a unique adaptation to tunnel living. When in the gallery, the wings are drained of hemolymph (the insect equivalent of blood), so they go floppy and are less cumbersome in the tight spaces. Outside, when preparing for takeoff, the web spinner pumps in fluid so the wings stiffen up enough to create lift.

Classification

ORDER	Embiodea
NUMBER OF FAMILIES	9
NUMBER OF SPECIES	360 approx.
DISTRIBUTION	Worldwide
HABITAT	Rocks and tree trunks in mostly tropical areas
NEST MATERIAL	Silk
DIET	Lichens and detritus

LEFT
IN THE GALLERY
Web spinners build galleries over their food sources, which are mostly bark, moss, and lichens. The females stay inside these tunnels for the duration of their lives.

CASE STUDY
Barklouse

Barklice scrape off the algae and fungi that grow on bark and grind it up. The name of the order to which the lice belong, Psocoptera, means "winged gnawers." Being only 5 mm long, most barklice are hard to spot, especially since they are often occupying damp nooks on tree trunks. However, the US species *Archipsocus nomas* is one of several gregarious barklice. They live in loosely associated social groups, all confined inside a thick web of silk.

FIRST GENERATION
Colonies of *Archipsocus nomas* are known in parts of the American South as "tree cattle." Their webs are rare over winter, but a few persist in oak "hammocks"—small woodlands—where the risk of frost is low. The first generation of barklice nymphs emerge in spring from eggs laid before the previous winter. Females are able to reproduce by parthenogenesis (they do not need their eggs to be fertilized by sperm). As such, populations swell rapidly, expanding protective webs as they go. By the early summer, large colonies become visible.

REPRODUCTIVE DIFFERENCES
During periods of parthenogenesis, males are not present in the colony, and females do not grow wings when they reach adulthood. During fall, once the colony has reached a large size, reproductive males and females are included in the later generations of eggs. These adults have wings, which means they can fly off to find mates in unrelated colonies and spread out to lay eggs on trees far from home.

ATTRACTION TO BOOKS
Smaller members of the Psocoptera, booklice are barely 2 mm long, mostly wingless, and survive on the starchy glues holding together old books—plus whatever else is growing on the aged paper. They are parthogenetic and multiply in number at alarming speed when they infiltrate a warm library. Old books are most at risk because modern volumes use less-palatable glues derived from petrochemicals.

Classification

ORDER	Psocoptera
FAMILY	Archipsocidae
SPECIES	*Archipsocus nomas*
DISTRIBUTION	Southeast United States
HABITAT	Damp woodlands and forests
NEST MATERIAL	Silk
DIET	Algae, lichen, fungi, and plant detritus

BARKLOUSE 59

ABOVE
DEFENSE CANOPY
The lice's thick, tangled web is a defense against predators and parasites. It can form a low canopy over the trunk, or be more tentlike when built around the union of several branches.

CASE STUDY
Acacia Thrips

Thrips are among the tiniest of insects, with adults being about 1 mm long. Roughly half of them eat fungi and a few are predators. The *Dunatothrips* of Australia are plant eaters, found only on species of *Acacia* plants with long, pinnate leaves. Members of this genus are semisocial, working together to build feeding and breeding nests by binding together the plants' leaves. Inside, the resident thrips work together.

THE DOMICILE
An *Acacia* nest is always founded by a group of thrips. Mostly this band comprises females, probably sisters, that have flown from their place of birth to find living space. They lash together the small, elongated leaves to create a sheltered space inside more easily. After construction is complete, the female builders lose their long wings so they can move around inside. Males flutter in to mate with the females. They keep their shorter wings while inside so they can set off to find more mates. Researchers have found that when a female is isolated from other thrips, she will not build a home for herself. This changes once she is joined by an adult male, although he never helps with the construction.

GENETIC DIFFERENCES
Females lay eggs in the nest. Thrips share the same system for determining sex as wasps, ants, and bees. Males develop from eggs that are unfertilized by sperm, and females come from fertilized eggs. The building of nests is strongly linked to males being around to mate, and therefore many—if not all—of the eggs laid in a nest hatch as more females. Thrips share the task of keeping the eggs clean, creating little garbage patches or middens for detrius. When the nest is too crowded, young sisters leave. They either build a new nest elsewhere on the plant or fly away in the hope of landing somewhere suitable for their new home. The rate of eggs laid that hatch into males is always low. Mature males always leave the nest. They may mate with their sisters living nearby or try their luck on the wind.

Classification

ORDER	Thysanoptera
FAMILY	Phlaeothripidae
GENUS	*Dunatothrips*
NUMBER OF SPECIES	7
DISTRIBUTION	Australia
HABITAT	*Acacia* trees and shrubs
NEST MATERIAL	Silk and leaves
DIET	Leaves

ABOVE
LIVING SPACE
The narrow fleshy leaves of the Acacia are bound together with silk at certain points. Inside, a community of tiny thrips consume the leaves' surface.

UNUSUAL NAMES

The term for a single member of the Thysanoptera order is "thrips," which is odd in many languages where a terminal "s" denotes a plural. To sow further confusion, the plural is "thrips" as well. The thrips' order name, Thysanoptera, means "fringed wings," and up close one can see that the wings have feathery edges. An insect of this size equipped with wings such as these is never going to be a strong flier and thrips are often blown enormous distances by wind, only to be dropped from the sky when the pressure drops. As a result, they are sometimes known as thunderbugs because they appear on crops after summer storms.

MATERIALS AND FEATURES
Gall Thrips

As well as building with silk, many thrips create galls on leaves and stems. Galls are growths that appear on plant leaves as a result of insects injecting chemicals into them. Thrips deliver these chemicals through their saliva while the tiny insects suck on the leaves. Often the chemicals make the leaf curl and thicken, thereby creating a safe space in which the thrips can feed and breed. Leaves with gall growths invariably die off earlier than they would otherwise.

In the case of species of *Kladothrips*, another Australian genus that targets *Acacia* trees, tiny fruit-like galls appear on the leaves. The galls provide shelter and are a source of food for generation after generation of thrips. Some species of *Kladothrips* have evolved complex social orders that are played out in the confines of the gall and which border on the sociality of wasps and ants in their complexity.

Fight for control

Galls are created by a founding generation of female *Kladothrips*. They limit the growth of the leaf, creating fleshy bulbs that resemble fruits by scraping away at the leaf's surface. The gall is produced by several females working on the leaf at the same time. However, only one can take ownership, and while the gall is taking shape, the founding females will fight for control. The victor lays her eggs inside the gall.

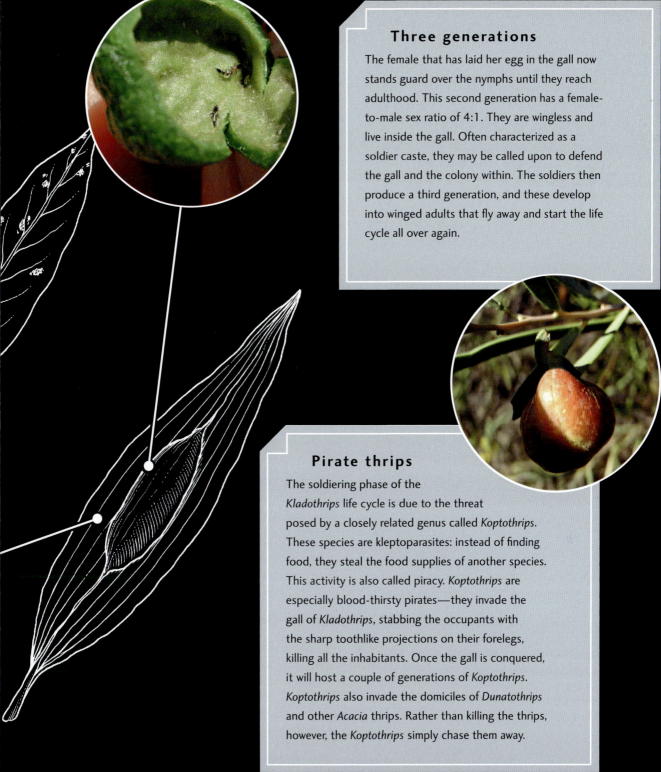

Three generations

The female that has laid her egg in the gall now stands guard over the nymphs until they reach adulthood. This second generation has a female-to-male sex ratio of 4:1. They are wingless and live inside the gall. Often characterized as a soldier caste, they may be called upon to defend the gall and the colony within. The soldiers then produce a third generation, and these develop into winged adults that fly away and start the life cycle all over again.

Pirate thrips

The soldiering phase of the *Kladothrips* life cycle is due to the threat posed by a closely related genus called *Koptothrips*. These species are kleptoparasites: instead of finding food, they steal the food supplies of another species. This activity is also called piracy. *Koptothrips* are especially blood-thirsty pirates—they invade the gall of *Kladothrips*, stabbing the occupants with the sharp toothlike projections on their forelegs, killing all the inhabitants. Once the gall is conquered, it will host a couple of generations of *Koptothrips*. *Koptothrips* also invade the domiciles of *Dunatothrips* and other *Acacia* thrips. Rather than killing the thrips, however, the *Koptothrips* simply chase them away.

CHAPTER THREE

Funnels, Cases, and Stalk Builders

As well as constructing places to live, breed, and raise young, some insect architects are experts in building structures used in defense and attack. For example, antlions build impregnable sand traps for capturing live ants. Glowworms—a particular kind from New Zealand, that is—lure flying prey into a sticky trap using a mesmerizing light. Meanwhile, green lacewing larvae clothe themselves in the skeletons of their victims. This practice builds up a suit of armor that to their prey looks, feels, and smells like one of their own but which obscures the killer within. And among the rock and gravel beds of clear rivers and streams, caddisflies are the insect masterbuilders of freshwater habitats, constructing funnels and tubes to filter foods from the water.

This chapter looks at the structures created by a disparate collection of insect orders that include true flies (Diptera), caddisflies (Trichoptera), and lacewings (Neuroptera), which also consists of antlions. All these insect orders are holometabolous, meaning they undergo a major transformation between larva and the sexually mature adult. In most cases, it is the larvae that are the builders. Larvae are focused on feeding and growing and, in the main, use the structures they create to catch prey.

TRUE FLIES

True flies are a well-known order, since they include the house fly and mosquito among their 150,000 species. The adults are two-winged insects known for their flying ability—the common name says everything. Ancestrally, flying insects have four wings, but true flies only use the two forewings for flight. Their hindwings, however, have not disappeared completely, and nor are they reduced into a supplementary adjunct flying surface as with hymenopterans such as wasps and bees, which are also superbly acrobatic fliers. Instead, the fly's hind wings are reduced to club-shaped structures called halteres. These are generally hard to spot in smaller species, but in larger and spindly creatures, such as crane flies, the halteres can be seen quite clearly poking out sideways from the thorax.

The halteres are free to move this way and that as the fly zigzags through the air. When the fly moves forward, the halteres swing back. When the fly rolls to the right, each haltere makes a movement in the opposite direction relative to the body. In this way the halteres are the fly's gyroscopic or

OPPOSITE
IT'S A TRAP!
These glowing droplets suspended on silk snare flying insects for a glowworm lurking above. The worm is the larva of a New Zealand fungus gnat (above).

orientation system and feed back to the fly a stream of information about its body position and motion, allowing it to pull off its acrobatic feats.

Adult flies mostly deploy three main feeding strategies and have mouthparts to match. Mosquitoes and gnats have sharp mouthparts that can pierce skin for sucking blood. House flies and fruit flies have spongelike mouthparts for slurping up liquid foods, be that the juice from a fruit or the ooze from a rotting carcass. It is the female flies that are drinking blood and doing most of the feeding. Unlike the males, they need nutritious meals to grow a supply of eggs.

Dipteran eggs mostly hatch into larvae that are better known as maggots. They are small and pale with no obvious delineation between head, thorax, and abdomen. Unlike in a beetle grub, for example, both ends of the maggot look rather similar. Looking much closer, the head end is generally narrower than the rear. The rear end may have a pair of dark spots, which might look like eyes at first glance. However, they are breathing holes.

There are a considerable number of exceptional larvae, and this chapter looks at a few in more detail. Others which are not included but deserve a brief mention here are black flies (Simuliidae; also called buffalo gnats) which complete their larval stage under water. The larvae use hooks on the ends of their abdomen to cling to a smooth rock on the bed of a fast-flowing stream, casting silken threads to other anchor points to hold them should the hooks fail. From there, they filter foods from the water with a brushlike mouthpart. After pupating, the winged adult needs to get out of the water, and it does this by holding a bubble of air around its body as it emerges from the pupal case, before rising to the surface in this escape capsule. By contrast, the alkali fly (*Ephydra hians*) of the hypersaline Mono Lake in California uses a similar technique to dive back into the water to feed on algae and lay eggs. The adults have waxy hairs that trap air around the body. Oxygen in the water passes into the bubble so the fly can breathe submerged for several minutes. In effect, the alkali fly builds itself an insect aqualung.

OTHER GROUPS

The life cycles of caddisflies resemble those of many dipterans that have an aquatic larval stage. However, caddisflies are more closely related to moths and butterflies. In fact, fluttering along on four wings with their long antennae in view, adult caddisflies are easily mistaken for moths at first glance. The difference is revealed in the names. Moths are lepidopterans, which means "scale wings." The oft-celebrated patterns on the wings of a butterfly or moth are made from fragile scales tiled over the surface. Meanwhile, caddisflies are trichopterans, or "hairy wings," and this refers to the fuzz of bristles on their wings. As we will see, the larval stages of a caddisfly can never be mistaken for those of a moth. They are almost exclusively aquatic, and involve some of the most intricate constructions of all insects.

The common name "lacewing" for members of the Neuroptera order is easy enough to grasp. The adults have long, near-transparent wings with a fine, complex venation. The scientific name for the order means "veined wings," because the framework of more rigid elements that hold the wing together are clear to see. The diaphanous quality of adult neuropterans belies the violence meted out by their larvae, which often use infernal constructions to trap and trick prey. They, like all insects in this chapter, use their architectural abilities to find food. Let's take a closer look.

OPPOSITE
CADDISFLY CASE
The larva of a diamond northern caddisfly constructs a sturdy case from flakes of wood.

BLUEPRINTS
Caddisfly Cases

Caddisflies are found close to freshwater habitats the world over. They are largely scavengers that eat the remains of dead water plants or scrape algae from the surface of rocks on the riverbed. Most caddisfly larvae are case makers, building a protective cocoon from materials gathered from their surroundings. The results are uniquely wonderful little edifices. And the presence of caddisfly is a signal that the ecosystem is thriving and healthy.

FIG. 1
FORM

The case is built from a wide range of materials. Wood fragments are commonly used for construction but so too are sand grains and fragments of the water plants, which the caddisfly often bites off for that specific purpose. The case is generally a hollow cylinder at least the length of the larva, and it is extended forward by adding new material to the opening at the front. The nature of the building material alters the overall look of the case.

The abdominal prolegs are limb-like appendages that protrude from the tip of the abdomen and have little hooks for gripping the inner wall of the case.

The building materials are glued together using a sticky silk that the larva produces from glands in the mouth.

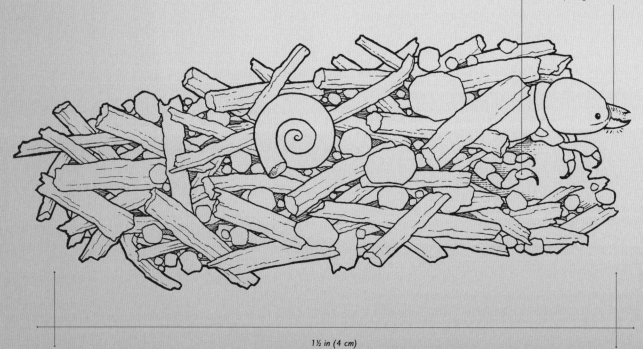

1½ in (4 cm)

FIG. 1 THE CASE FORM

PUPATION

Assuming they get enough food to grow to full size, caddisflies will use their cases for pupation. The pupa is active: it moves to a safe location, ties down the case, and blocks up the entrance before becoming dormant as it completes development. The pupa then bites its way out of the case and floats or crawls up to the surface. Floating at the surface or on the bank, the pupa's body then splits lengthwise, releasing the winged adult within.

By using materials collected from its immediate surroundings, the larva is able to hide in plain sight, as its case is hard to make out from the substrate upon which it sits.

Retreating caddisfly are unfussy eaters and sift silts that flow through the tubular cases for food.

FIG. 2
FUNCTION
There is an opening at the rear of the case, smaller than the one at the front, which allows a steady current of water to flow through. This flow is encouraged by the caddisfly wriggling its body, and ensures that a good supply of fresh, well-oxygenated water reaches the animal's gills. Some caddisflies give away their position by poking their heads and thoraxes out of the case and hauling it across the riverbed, to look for food or graze on algae. When disturbed, the larva darts back inside for safety. A safer strategy is for the larva to anchor its case against the current, lashing it to a stem or gluing it to a rock. This form of case is known as a retreat, and the insect lurks inside waiting for a morsel of food to float past.

FIG. 2 THE CASE FUNCTION

CASE STUDY
Net-Making Caddisfly

The common name "caddisfly" is thought to come from "caddice," a Middle English word for silks and other fine fabrics. Caddice men were traveling salesmen who pinned samples of their best cloth to their clothes. Some caddisfly in the suborder Annulipalpia (and a few other families), do something similar: they weave a finely meshed net of silk to filter food fragments from the water.

Classification

ORDER	Trichoptera
SUBORDER	Annulipalpia
SPECIES	4,600
DISTRIBUTION	Eastern North America
HABITAT	Streams
NEST MATERIAL	Silk
DIET	Organic particles in water

NET STRUCTURE

A net-making caddisfly generally lives in shallow streams that have pools of slower water. It is found in both clear waters flowing through rocks and those muddied by clays and silt. The insect builds its net in a space beneath a rock, where the water current funnels through it. The nets of *Dolophilodes distinctus* are the finest woven by any caddisfly known, in that the openings of its mesh are about 5 microns across. The larva can spin around seventy strands at a time, and will enlarge its net as it grows. A final-instar larva, one that will soon be ready to pupate, maintains a net that is just over 2 inches (6 cm) wide at the front end and up to 6 inches (15 cm) long. A net like that requires $2/3$ mile (1 km) of silk strands and has 100 million openings.

COLLECTING FOOD

Water is able to flow through a caddisfly's net, but suspended organic particles are snared. These particles are mostly microscopic algae, such as diatoms. The larva does not bother to build a solid case. The net is its retreat, and it spends its days climbing around inside eating whatever edible particles it finds and clearing away the sediments to maintain the net's efficacy.

NET-MAKING CADDISFLY 71

ABOVE
RICH PICKINGS
Dozens of Neureclipsis bimaculata have constructed nets across a stretch of clear and steadily flowing water.

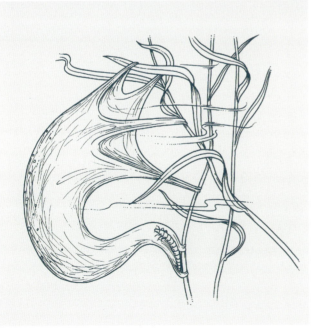

LEFT
TRUMPET SHAPE
The shape of the net of members of the Polycentropodidae family has earned them the common name of trumpet-net caddisflies.

BRANCHING OUT

Species of the Dipseudopsidae family have innovated their nets so that they are buried in the sediment and branch off in several directions, curving this way and that under the stream bed. An upstream opening above this sediment scoops in a flow of water which leaves via a single outflow downstream. Inside its underground network, the larva spins webs across the passageways to collect particles that flow through.

MATERIALS AND FEATURES
New Zealand Glowworm

The deep caves of New Zealand glow with an eerie blue-green light. The light comes from droplets of sticky mucus that dangle from the ceiling by strands of silk. This natural chandelier is not there to help insects fluttering lost in the dark, but is intended to snare them, so that its creator, the wormlike larva of a fungus gnat, a member of Diptera, can slither over and eat them. The gnat species (*Arachnocampa luminosa*) is better known in New Zealand as a glowworm. However, it should not be mistaken for the more widespread glowworm, which is also not a worm but the attention-seeking wingless females of a family of beetles. (Fireflies are also beetles and have the added ability to glow while they fly.) The dipteran glowworm glows thanks to a luciferase (an enzyme that generates light). The chemical is produced in its Malpighian tubules (an excretory system). The luciferase is then used to oxidize another substance called luciferin, and that chemical reaction releases light. This is the process used by other bioluminescent insects, including glowworms, but *A. luminosa*'s enzyme has a unique structure, suggesting this is a clear example of convergent evolution.

Hanging around

The glowworm hangs from the roofs of caves, lying on its back within a hammock-style tube of silk threads, known as its gallery, which is stuck to the rock above with anchor lines. The gallery needs to be two or three times the length of the larva. The larva continually tends to the threads, adding fresh droplets and more silk. The first droplet produced by the larva is always the largest. Its weight keeps the thread taut and enables the larva to make it longer by adding silk at the top. The larva does this by dangling the first third of its body down from the gallery. Its mucus then accumulates into droplets as it spins out more silk.

Setting the traps

The glowworm steadily progresses across the cave roof, paying out the strands of a gallery structure that holds its weight and acts as scaffold for its feeding snares. The gallery has a curtain of up to seventy silk threads, all coated in glowing sticky mucus that is doped with activated luciferin. The larva moves back and forth tending to its traps, but it cannot slither backward, only forward. It has tiny bristlelike projections on its underside that would rip the gallery if it went in reverse. So the larva must turn around in the gallery to change direction.

Feeding time

In the dark of the cave, the light show is irresistible, its purpose to ensnare flying insects. The curious prey gets stuck to the mucus, and as it struggles to escape, it only becomes entangled further. To retrieve its meal, the glowworm stretches out of the gallery and hangs down beside the snare line. It then hoists the snare upward by shortening its body into a thicker form. It can haul up around 3 mm of line with each heave. The larva has powerful biting mouthparts that subdue the prey, which is then sucked dry. Larger prey items are left hanging so the larva can eat the leftovers later.

Building an Antlion Pitfall Trap

As an adult, an antlion (from the family Myrmeleontidae) resembles a damselfly, buzzing around slowly on four narrow wings. The graceful adults are active at night and live for just a few weeks at the height of summer as the females scatter their eggs over sandy ground. The rounded larva that emerges from the egg has a very different demeanor. Also called a doodlebug, it is seldom seen aboveground. A fierce predator of ants and other ground insects, it builds conical sand traps angled to collapse if anything walks on them, causing its prey to tumble down into the middle, where the gaping maw of the antlion is waiting.

3 A downward spiral ↓
Still under the surface, the antlion spirals inwards from the trench, flinging out sand at it goes to create a deeper crater. Once it reaches the middle of the circle, it digs down to deepen the crater into a cone shape. It keeps on digging until the side of the slopes approach 45 degrees. At this angle the sand is on the verge of collapse.

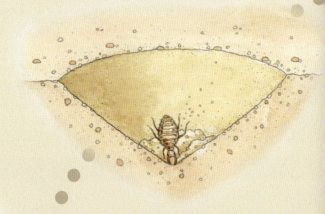

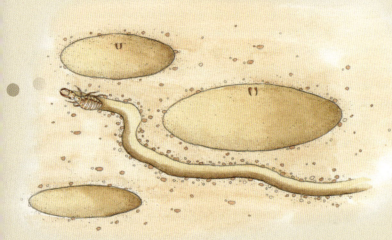

BELOW
DEADLY BITE
At the base of the trap, the antlion's long, pincer-shaped mouthparts are ever ready to strike.

1 The search for sand
After hatching, the antlion larva hunts for a suitable place to make a nest, which also serves as a trap. They need a patch of fine sand that is at least a couple of inches deep. The trap also needs to be set up in a sheltered spot so the wind and rain do not damage its carefully constructed sandpit.

2 Draw a circle ↑
Once the larva has found a suitable location, it burrows below ground and shuffles backward in a circle. By flinging the sand it displaces up and outward, it creates a circular trench in the sand. A large, late-stage antlion larva digs trenches that are just over 2 inches (6 cm) across, but the younger ones start smaller.

BUILDING AN ANTLION PITFALL TRAP

4 Springing the trap →
The antlion positions itself at the center of the trap with the tips of its pincerlike mandibles just poking out. When an ant (or another small ground creature) wanders into the trap, the sand underfoot slides down the slope. The ant follows. The antlion is ready to pounce and will fling the ant around until it is positioned correctly for a killer bite. Try as it might, the ant cannot escape. The antlion will throw loose sand at the ant to dislodge the slope as it tries to climb out.

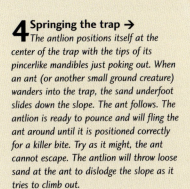

5 Venomous bite ←
Once the antlion captures its prey in its mandibles, it injects a fast-acting venom that paralyzes the victim. The antlion then sucks out the body fluids of its prey, leaving a dried husk.

6 Slow development →
An antlion larva often has to wait for weeks for its next meal and so it grows only slowly. It can take three years for the larva to reach a size suitable for pupation, which happens in spring. The cocoon is buried at the bottom of the pit, and the adult emerges by summer, which gives it enough time to mate and lay eggs before fall. The larvae hatch in fall and spend their first winter dormant in a deep burrow.

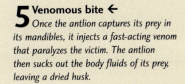

CASE STUDY
Wormlion

The wormlion is so named because, like its distant cousin, the antlion, it builds pitfall traps to catch meals. However, while antlions devour ants, wormlions do not target worms. They too are ant eaters, and their name really refers to the fact that their larvae, as they build their traps, resemble wormy maggots. The larvae take this form because they come from a family of true flies—a highly unusual one called the Vermileonidae.

CONSTRUCTION METHOD
Once complete, the trap of a wormlion looks more or less indistinguishable from that of an antlion, and it functions in the same way. However, the wormlion constructs its trap using a much more laborious and time-consuming method. The larva starts work in the middle of the pit with its rear half buried in the sand, and the tip of its abdomen, which has four protruding anchors, holding it in place. It then flexes the upper end of its body, digging its head into loose sand and flinging it outward by straightening its body. As the pit deepens, the flicked sand trickles down the sides, creating the unstable slope needed to trap prey.

IN THE PIT
When the trap is set, the wormlion writhes in the sand to coat itself in a thin layer of grains. This is sufficient camouflage for it to sit in the base of the pit and wait for a victim to tumble in. When that happens, the wormlion undermines escape attempts by flinging more sand at its prey as it tries to climb out. The result is a steady slide of sand that always brings the prey back to the bottom of the pit. There the wormlion holds it down with the help of a little pseudopod, or temporary, appendage on its fifth body segment. Its bite then injects a venom and digestive enzymes that turn the prey's insides to mush.

Classification

ORDER	Diptera
FAMILY	Vermileonidae
NUMBER OF SPECIES	30 approx.
DISTRIBUTION	Worldwide but mostly in Africa
HABITAT	Sandy areas
NEST MATERIAL	Sand
DIET	Ants and other small ground insects

**LEFT
HOLD ON**
The stumpy appendages poking out from the abdomen act as anchors to hold the larva in place.

**ABOVE
GATHERING**
The wormlions tend to gather in areas where the sandy soils are suitable for their traps.

ENEMY ATTACK

Several types of ground-hunting birds have learned to recognize the pockmarked surface areas of wormlion pits as a source of food. When in danger of being eaten, the wormlion buries itself in the pit, taking on a curved posture to make it harder for a bird to tease it out. If that proves unsuccessful, the wormlion writhes powerfully, often exerting enough force to fling itself to safety.

CASE STUDY
Pearly Green Lacewing

A female green lacewing is an egg-laying machine. Her offspring feast on the aphids and other bugs that infest plants over summer, and so she will be ready to lay eggs around June. Over the course of the summer she will lay around three hundred eggs, fueling her mission with meals of sweet nectar, pollen, and aphids. The combined weight of her eggs is twice her own body weight.

A NEST OF STALKS

The female lays around ten eggs at a time, mostly at night. She creates a little nursery on a plant that is infested with aphids. Where there are aphids, there are also ants, and they represent the largest threat to the lacewing eggs. To protect her eggs, the female creates a sticky blob of liquid silk with glands linked to her genital and egg-making organ. She dabs that blob on the underside of a leaf, and draws it into a thin stalk about 1/3 inch (1 cm) long. Once the silk is at full stretch, she seamlessly attaches an egg to its tip. After a wait of a few seconds to let the silk crystallize, she lets go, and the egg is held in place by the stiffened stalk. The silken stalk is so fine that it goes unnoticed by any ants that march across the leaf.

WINTER COLORS

The life cycle of the pearly green lacewing, so called because of its large glossy eye, is typical of other green lacewings. The adults need to mate and start laying eggs by early summer to exploit the aphid food supply which explodes around them, and that means the adults emerge from pupation in fall and must find a place to overwinter. The pearly green lacewing hunkers down in leaf litter, and just as its green coloring helps it blend in during the summer, the adult forms of these insects change to brown so they can hide among fallen leaves.

Classification

ORDER	Neuroptera
FAMILY	Chrysopidae
SUBFAMILY	Chrysopinae
SPECIES	*Chrysopa perla*
DISTRIBUTION	Europe and northern Asia
HABITAT	Grasslands and woodland edges
NEST MATERIAL	Silk
DIET	Nectar and aphids

PEARLY GREEN LACEWING

LISTENING IN

Hibernation is over by May, at which point the young adults need to find mates. The lacewing does this by sending vibrations through leaves and stems, which it does by standing on the plant and shaking its abdomen. The motion is transmitted through the plant to another lacewing nearby, which can feel the mating call with its feet.

ABOVE
STIFF STUFF
Lacewing silk has a rare combination of cross-linkages between the proteins, which makes it stiff.

CASE STUDY
Common Green Lacewing

The larvae of common green lacewings are voracious predators from the moment they hatch. It takes about four days for the larvae to be ready to leave the egg, which is when the tiny hunters clamber along the stalk to the leaf and start looking for food. Despite having pincerlike mouthparts that both grab and inject a paralyzing venom into prey, these weapons alone are not sufficient. And so, the larvae set about creating an extraordinary disguise.

EXOSKELETON OUTFIT

The lacewing's venom is a concoction of many substances. As well as paralyzing aphids and other prey, the venom turns the victim's insides into a liquid that the lacewing can then suck up and digest. What is left is an empty husk of exoskeleton. Rather than toss this away, however, the larva glues it to its back with a dab of sticky silk. By itself, this adornment is not particularly useful, but after a few meals, soon the whole larva is covered in a haphazard outfit constructed from multiple exoskeletons plus scraps of leaf and lichen. For this reason lacewing larvae are sometimes called junk bugs, and yet the suit has a clear purpose, fooling any ants nearby into thinking that the hunter looks and smells like a harmless aphid and quelling their urge to attack interlopers.

TWO SILKS

Lacewing larvae have no rear vent for expelling feces; that waste is stored in the body, in the same spot as where their silk glands are located. These are used to glue a larva's camouflage jacket together and to spin a cocoon when the time comes to pupate in late summer. The purpose of the cocoon silk is to act as a framework for an inner lining of fats that waterproofs the pupal chamber. The silk used is twice as thick and stickier than the threads used by the adult female to make egg stalks. She uses a genital gland, while the larvae use a gland in the excretory Malpighian tubules, to produce silk. This is a rare example of an insect that can produce two kinds of silk, albeit at different times of life.

Classification

ORDER	Neuroptera
FAMILY	Chrysopidae
SUBFAMILY	Chrysopinae
SPECIES	*Chrysoperla carnea*
DISTRIBUTION	Northern America, Europe, and northern Asia
HABITAT	Woodlands and meadows
NEST MATERIAL	Silk and animal bodies
DIET	Aphids and other small insects

COMMON GREEN LACEWING   81

ABOVE
TROPHY CLOAK
The young lacewing larvae build a camouflage cloak, which they make from the bodies of their victims.

CHAPTER FOUR

Wasps

Ask anyone "What is a wasp?" and they'll likely tell you some version of "a flying insect with yellow and black stripes and a nasty sting." They may even describe their papery nests, hidden away in roofs or trees. However, this is only a fraction of the story. In fact, most species of wasp do not sting, many build smaller nests from mud, but almost all live a life with no homes at all.

The wasps belong to the insect order Hymenoptera, which also includes bees and ants, and thus forms a significant community among insects that build structures. (All ants and bees are actually special kinds of wasp.) The order is named for the way these insects' two sets of wings are hooked together to make a single membranous flight surface (the hindwings are much the smaller). Almost all members also have the familiar narrow "waist" between the thorax and abdomen.

There are at least 150,000 species of wasp. The majority of them are carnivorous, eating other animals, and most are parasites—or, more accurately, parasitoids.

A parasite is an organism that takes resources from an unrelated host species. The host may not die, but is definitely weakened by this association. A parasitoid, meanwhile, is somewhere between a parasite and a hunter. In terms of wasps, the female seeks out a very particular prey—it could be a specific caterpillar, spider, or beetle grub—but she does not kill it. Instead, she lays an egg on it, or inside it, using a long, spiked ovipositor, or egg-laying organ, that is plunged into the victim's body. (Some are sharp enough to penetrate wood, and it is this feature that is repurposed as a sting in the Aculeata—the stinging Hymenoptera.) After the attack, the victim is largely unfazed and returns to life as normal. However, once the egg hatches, the host becomes a living, breathing larder for the wasp larva. It eats the host alive from the inside out, and then pupates inside the hollowed carcass. The adult wasp that emerges generally survives on pollen and nectar as it searches for a mate, before the macabre hunt for fresh meat begins again.

Nest-building wasps evolved from their parasitoid ancestors, and have retained the strategy of provisioning their larvae with food, though instead of laying their larvae inside their prey, they create secure nurseries for them, which they then stock with food.

Wasps are generally solitary creatures, and builders choose a secluded spot for their nesting structures. Others form small communities, building their nests in clusters, sharing the labor but keeping each larval nursery separate. Yellowjackets and their larger cousins, hornets, by contrast, are eusosocial, forming highly cooperative groups where only a single queen is permitted to lay eggs. These hatch mostly into daughters, which then aid their mother in producing more and more young. They do this in part by constructing elaborate nests that house hundreds upon hundreds of larvae, each one feasting on food delivered by their sisters.

RIGHT
HOVER-WASP NEST
The base of this nest, built by Eustenogaster calyptodoma, has a small spoutlike opening.

BELOW
CLAY LARDER
An excavated mud nest of a potter wasp shows cells filled with green caterpillars and yellow larvae in various stages of development.

BLUEPRINTS
Building Materials

Nest-building wasps, both solitary and social, prefer to make their nests in dry places where fungus and molds cannot attack the young developing inside. As such, green plant materials are seldom used in nest-building. Instead, the wasps lean on two dry materials: earth and wood.

**FIG. 1
USING WOOD**

Building with mud is a costly business, requiring a lot of water and labor. Additionally, the nests are heavy and need to be located in a spot with adequate support. Eusocial wasps, such as yellowjackets, hornets, and hover wasps, have evolved to build with a more efficient and less costly building material: a natural paper manufactured from wood fibers. This material is lightweight, strong, and plentiful. Much faster to build with and requiring less water to create, it is no surprise that all eusocial wasps use the material to build their intricate nests.

Almost uniquely among wasps, the males stand guard as the nests are constructed, wary that a parasitic wasp might sneak her egg into a cell in place of his own offspring.

Polistes wasps are known as umbrella wasps due to the way their paper nests resemble an upside-down umbrella.

Paper-building wasps have evolved a range of galleries and combs of cells for brooding larvae. The ultimate goal of the nest is to produce a new crop of queens and some male suitors. These leave home at the end of summer to mate.

Wasps, in common with all hymenopterans, have chewing mandibles, and paper wasps use them to harvest fibers for their paper. Production begins as the wasp scrapes fibers from dry wood or plant stems. The fibers are mixed with saliva to make a pulp, which is spread into sheets that dry into a solid material.

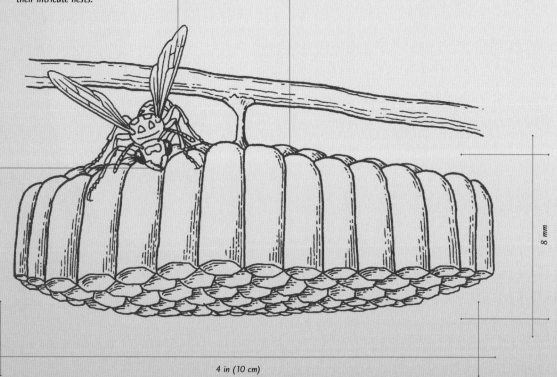

8 mm

4 in (10 cm)

FIG. 1 PAPER NEST

BUILDING MATERIALS 85

**FIG. 2
IN THE EARTH**

Several solitary wasps build tunnel nests. The digger wasps (Crabronidae) are named precisely for this reason. They tend to burrow into sand or dry earth to create simple spaces for one larva to develop. Other earth-builders, such as the potter wasps and mud-daubers, construct nests aboveground, on branches, under leaves, or on rocks and walls. Mud-daubers (a relative of the diggers) rely on damper habitats where they can find a supply of moist clays which go hard when they dry out. Potter wasps (also called mason wasps; Eumeninae) live in drier places. They take a big drink of water and then regurgitate over some dry soil to create a ball of mud that's good for building.

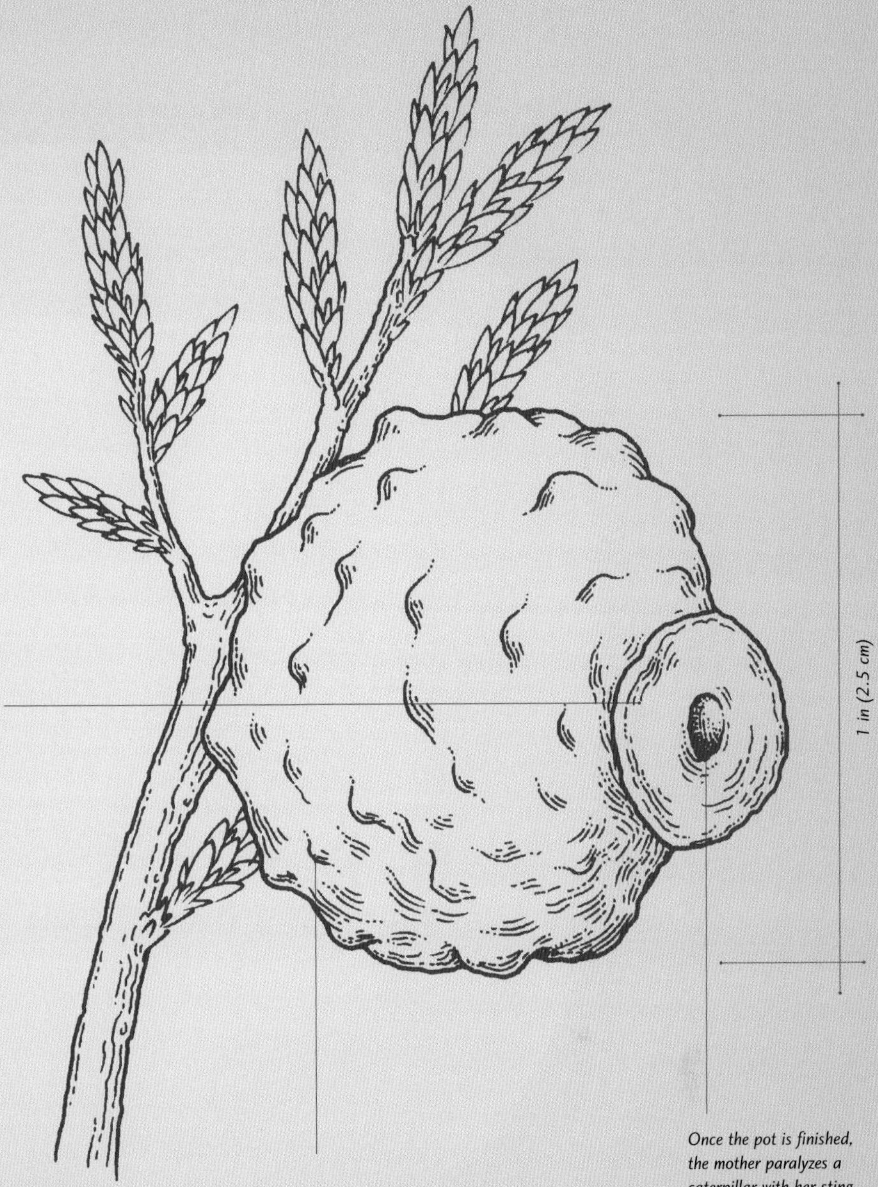

The female potter wasp collects clay and builds a flask-shaped nest for her young. There is one entry point forming the neck of the flask.

1 in (2.5 cm)

When one nest is completed and filled, the female wasp seals the entrance with damp spoil then sets off to build another. She feeds on pollen and nectar to fuel her activity.

The nest is painstakingly constructed from balls of clay collected nearby and wettened with saliva. It is attached to a plant stem or rocky surface and hardens once dry.

Once the pot is finished, the mother paralyzes a caterpillar with her sting and hauls it back to the nest. She will carry this fleshy prey into the flask and lay an egg on it. In this way, when the egg hatches, the larva will have plenty to eat.

FIG. 2 MUD NEST

Life Cycle of a Common Wasp

The common wasp *Vespula vulgaris* is also called the Eurasian yellowjacket, even though it has now found its way to Australia and New Zealand. (It was once thought to be at home in North America too, but the common wasp there has recently been designated as a separate species.) Starting in spring, and over a period of around five months, a single common wasp queen can produce a nest of seven thousand workers—just the resources she needs to produce a new generation of queens. They will mate in late summer and hibernate through the coming winter before building their own nests the following spring.

2 Work begins ↘
The queen busily fetches pulp to create a spindle, a sturdy anchor point from which the whole nest hangs. This element is maintained and strengthened throughout the life of the nest as it grows. The spindle—also called the petiole as it extends through the nest—is a vulnerable point and so is coated with repellant chemicals that deter ants.

1 Waking up ↓
After emerging from hibernation, the queen feeds on early blooms of nectar and pollen and then searches for a nesting site. The location needs to be frost-free and not too hot and dry; young queens often set up in outbuildings that fit these requirements. The site also requires a solid and reliable anchor point, which could be a root in an underground hollow, a branch of a tree, or a beam in a quiet roof space.

3 First gallery ↓
The queen builds a comb of hexagonal brood cells for her eggs. To grow the comb, the queen crosses each time to the opposite side to add the next cell, thus building them in a spiral order. She also builds an umbrella-shaped envelope over this first comb as protection. There is a wide opening beneath to allow plenty of room for expansion.

BELOW
ENTRANCE
Underground wasp nests are accessed through a single entrance tunnel. Underground, the nests hang from a sturdy root. Aboveground, they hang from branches or roof beams.

LIFE CYCLE OF A COMMON WASP

4 Laying eggs ↑
When the first comb gets large enough—normally between about 20 and 30 cells—the queen starts to lay eggs in each cell. The queen sustains herself on pollen and nectar and feeds her first generation of larvae on chunks of caterpillar and other prey. Common wasps kill with their biting mouthparts. The prey is then drained of liquid and butchered into chunks that are transported to the nest and used to stock each brood cell.

5 First workers
The first comb of larvae pupate into adult workers about thirty days after the founding of the nest. They take over building successive galleries of brood cells, one below the other. The queen has now lost the ability to fly and focuses on egg laying. She spends the next few weeks laying around 250 eggs a day, with more and more workers emerging each day to care for them.

6 Full size ↗
By the third month, the nest is reaching its full size. It has several galleries now shrouded in a flask-shaped envelope open at the base. Larger nests have thinner envelopes, because they produce more heat and so need less insulation. The queen's rate of egg laying has slowed, so the colony nears a replacement rate, where the number of young workers emerging each day matches the number of older dying. The colony is no longer growing in numbers.

7 New queens
In the fourth month, the workers build queen cells where young queens and drones (reproductive males) are to be brooded. The workers divert most of the food to these cells. Remaining worker larvae will probably starve. Once the virgin queens and drones have left to begin the cycle afresh, the queen dies.

8 The end
The colony begins to collapse and workers start to lay their own eggs, cannibalizing their sisters to provision the cells. These newly reproductive workers cut away cells from the nest and fly away instinctively, attempting to set up a new colony. However, any eggs that do hatch are always males incapable of building nests. As the wasp nest shrinks, it becomes cooler. Eventually, any wasps that remain succumb to the winter cold.

MATERIALS AND FEATURES
Gall Wasps

Gall wasps (from the family Cynipidae), sometimes better known as gallflies, are minute members of the order Hymenoptera. Some are only 1 mm long, and it is reckoned that there are about 1,500 species of these solitary wasps. Some are exclusively parthenogenetic, meaning they can reproduce asexually without the need for eggs to be fertilized by sperm, but the majority have a sexual phase, switching to parthenogenesis once a year.

The tiny females lay their eggs in plant tissue, generally in the leaf or a twig, and each species targets a particular plant host. Once the egg hatches, the wasp larva secretes a cocktail of chemicals that mimic and manipulate the growth hormones in the plant. The exact process is still somewhat obscure, but the larvae are able to induce the plant to create a fleshy growth called a gall by hijacking a leaf bud or branching twig. The galls, which appear in a range of shapes and locations, function as a protective nest that also provides a food supply for the tiny larva as it develops into an adult.

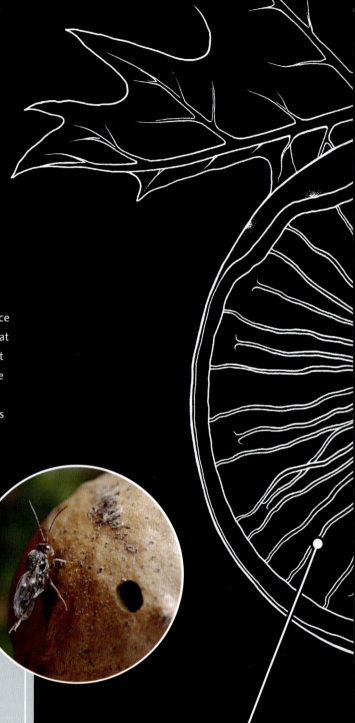

Some apple galls are filled with a solid flesh much like their namesake. This one is largely hollow save for a spray of supportive filaments. The wasp larva occupies the fleshy central hub where it eventually pupates. Once the new adult has emerged and punched its way out, the hollow galls dry out and wither quickly.

Wasp eat wasp

The ecology of gall wasps is enormously complicated. Many gall wasps are actually parasites of the gall-making species. The females sync their egg laying with the growth of particular galls so their larvae can benefit from the gall's nutrition as well. They are inquilines or "tenant" species that do not hinder the "landlord" species. The same is not true of the parasitoid gall wasps that target galls specifically for the plump larvae inside. And yet other hyperparasitoid species will arrive later to target those invaders! Despite the turmoil that may or may not be happening inside, however, the galls are largely harmless to the plants.

GALL WASPS   89

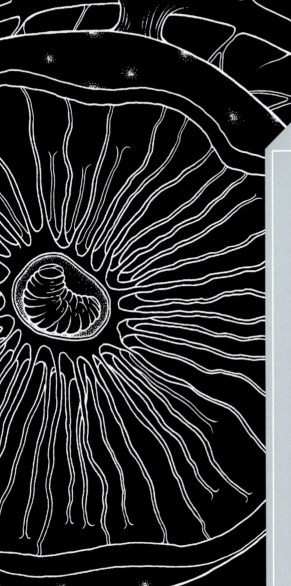

Types of gall

Galls are often rounded and easily mistaken for fruits such as berries or even crab apples.

Cherry gall: Turning from green to bright red, these galls resemble the cherry in size and color but grow on the underside of oak leaves.

Apple gall: Another oak gall, these look like little apples (being up to 2 inches/5 cm across) and swell out from the twigs of oak trees.

Knopper gall: These knobbly amorphous green galls develop on the acorns of certain oak trees.

Pincushion gall: Many galls have a spiked or feathery appearance. Robin's pincushion galls are seen in wild roses (right).

Marble gall: These are like apple galls, sprouting from twigs, but are more rounded and are rarely more than 3/4 inch (2 cm) wide.

Spangle gall: These are flattened disk-shaped galls, measuring no more than 5 mm. They grow under leaves. There are often hundreds on a leaf, each one with a single wasp larva inside (left).

CASE STUDY
Pollen Wasp

Pollen wasps (Masarinae) will sting for defensive purposes—and theirs is a painful one at that—but not to kill prey. The short-lived males barely eat at all, and die after mating. The females take small meals of pollen and nectar, both high-energy foods. The females also collect pollen grains soaked in nectar from summer flowers and use that to provision a handful of cells in a small mud nest.

Classification

ORDER	Hymenoptera
FAMILY	Vespidae
SUBFAMILY	Masarinae
DISTRIBUTION	Southern Africa, North and South America
HABITAT	Desert
NEST MATERIAL	Mud
DIET	Pollen and nectar

BUILDING THE NEST
Pollen wasps locate their nests out of sight, perhaps under a rock or in a crevice in a high cliff or wall. Less commonly, the wasps build directly on the ground or on tree branches. Females are the only builders. They gather soil in their mouthparts and mix in saliva to create a ball of clay. They create dozens of these balls to construct tubular cells, lined up side by side. The mud walls need to be thick enough to prevent predators from getting in—not least parasitoid wasps that might try to jab their ovipositors through.

FOOD SUPPLY
The lifestyle of a pollen wasp is superficially similar to that of a honey bee, since both collect food from flowers. Yet that is where the similarity ends. The pollen wasp eats the pollen grains and nectar and flies back with it in its crop, a sac at the top of its digestive tract. The wasp then regurgitates a slurry of nectar and pollen into each cell and lays an egg in this mixture before sealing the cell shut with more mud.

MUD NURSERY
The larva hatching from the egg must rely on the pollen supplies left by its mother. Once full size, the larva prepares to pupate by spinning a silken cocoon that lines the inside of the cell. Once transformed to an adult, the pollen wasp gnaws its way out through the dried mud entrance to the cell. The nests are sturdy enough to persist for the rest of the summer, and are often found with the cells open and the silk remnants of the cocoon still visible.

POLLEN WASP 91

ABOVE
CLAY NEST
The mud nest of the pollen wasp Pseudomasaris vespoides *is glued to the side of a house. This wasp lives in the western United States and can be seen on beardtongue plants.*

CASE STUDY
Potter Wasp

Potter wasps (Eumeninae) belong to the same wasp family, Vespidae, as yellowjackets (Vespinae), paper wasps (Polistinae), and other eusocial wasps that build large nests for large colonies. However, potter wasps are solitary and work alone. The females sculpt a home for their broods. Some take over pre-existing holes, even those made by a long-gone nail, or build a house from chewed plant fibers, making a rough kind of paper.

MUD BALL

Adult *Eumenes* potter wasps drink nectar as they search for mates. Once mated, the female seeks out a patch of soft, fine-grained clay and begins to excavate it with her mouthparts. She regurgitates water into the soil to soften it into mud and begins to build a ball with her front legs. When the ball is around the same size as her head, she flies off with it to the nesting site. This can be on a rock crevice, a tree trunk, or a leaf or stem that is strong enough to support the brood cell. Several mud balls are needed to build a near-spherical nest with a small opening (around 2 mm across) at the front. In the space of a few months, a female can build around twenty of these "pots"—sometimes separately, sometimes in clusters.

JUGGED FOOD

Once the mud house is completed, the mother lays a single egg inside. Next she seeks out something for her offspring to eat once it hatches. Usually this will be a juicy caterpillar, which she has paralyzed with venom and packed away inside the jug-shaped nest. Some species lay the egg after provisioning the nest, dangling it on a thread of silk at the entrance. When the egg hatches it consumes the caterpillar, taking anything from a few weeks to a whole year to complete its development. That period is generally slower in cooler habitats and synchronized with the changes in seasons. The nest is sealed after provisioning, but there is always a danger that a cuckoo wasp has laid its egg inside during its construction. The larva of this brood parasite hatches first then eats the caterpillar and the potter wasp larva in turn.

Classification

ORDER	Hymenoptera
FAMILY	Vespidae
SUBFAMILY	Eumeninae
DISTRIBUTION	Worldwide
HABITAT	Warm and tropical areas
NEST MATERIAL	Mud
DIET	Caterpillars and nectar

POTTER WASP 93

LEFT
INSPIRING NEST
Members of the genus Eumenes *are typical of the subfamily in that they build a jug-shaped nest for their larvae—and perhaps inspired Native American potters with their designs.*

CASE STUDY
Hover Wasp

Hover wasps, a little known subfamily in the Vespidae, represent a fascinating half-way house between solitary potter wasps and highly social paper-building wasps. Hover wasps set up small colonies in mud and paper nests with irregular shapes. Each nest is built by one or two female foundresses, both ready to lay eggs. The first generation of eggs comprises almost exclusively females, who help with further construction, and one day may inherit the nest.

NEST FOUNDATION
The hover wasp foundress hangs her nest from a branch and builds a series of brood cells. She lays an egg in each one and coats it in a general secretion from the Dufour's gland (named after naturalist Leon Dufour who first described it in 1841) in her abdomen. This gland is the same that contains the sting venom in social wasps and others. While hover wasps can sting, they tend not to. Instead, the gland secretions are used by newly hatched larvae as a medium for receiving nutrients. Adult carers—the foundress mother at first, but later it could be an older sister—regurgitate food onto the gel-like mass, and that makes it easier for the larva to eat. The adults are predators, killing caterpillars and other prey. (They are also known for stealing food trapped in spiders' webs.)

STAY OR GO?
The foundress cannot lay eggs indefinitely, and runs out far sooner than a yellowjacket queen, for example. A hover wasp colony may only grow to include ten or so members, but there have been examples of larger groups. The colony size is limited by the motivations of the foundress's daughters, which, unlike eusocial workers, are all fertile. They face a quandary as soon as they emerge as adults: do they leave and risk founding their own nest, or should they stay on to help in the hope of being able to take over when the foundress dies? That decision depends on how many daughters are currently working in the nest; with every new generation there is increasing competition, and so the option to leave is the smarter one.

Classification

ORDER	Hymenoptera
FAMILY	Vespidae
SUBFAMILY	Stenogastrinae
DISTRIBUTION	Indo-Pacific region
HABITAT	Tropical forests
NEST MATERIAL	Mixture of wood fibers and mud
DIET	Insects

HOVER WASP 95

LEFT
VASE SHAPED
The narrow entrance spout at the base of this paper nest gives it a shape typically associated with hover-wasp nests. There are normally a dozen or so brood cells inside.

CASE STUDY
European Hornet

The hornets of the genus *Vespa* are the largest eusocial wasps of all. They can be more than 2 inches (5 cm) long and their stings are more painful than those of smaller wasps, some delivering a potent toxin. The Asian giant hornet (*V. mandarinia*) is especially venomous. In Japan it kills around forty people a year. (True hornets live mostly in Asia and Europe. Often the insects known in America as hornets are in fact species of yellowjackets.)

CYLINDRICAL NESTS

European-hornet nests are typical of *Vespa* species in that they are broadly cylindrical, not a ball like those of the closely related yellowjackets. They hang from a sturdy branch, and are also shielded by an outer envelope so the combs inside are shrouded in darkness. The nests are mostly built in secluded, gloomy spots, such as tree hollows, to further reduce its exposure to light. (Asian giant hornets are unusual in that they build nests almost exclusively under the ground.) Finally, the big giveaway that this is a hornet's nest is that it has a single, wide entry hole at the bottom, with little attempt to seal in the combs inside.

BUILDING MATERIALS

The nest is made from fibers of chewed dead leaves, bark, or wood, but often a mixture of all three. This is made into a pulp by the workers' saliva. When building, the sheets of pulp are further coated with saliva to help them stick together and to add a layer of weatherproofing, keeping out the wind and rain. A high proportion of saliva in the nest wall helps it absorb water and prevents damp problems inside. Analysis also shows that the hornets add traces of soil to the pulp, too, as evidenced by the presence of metallic compounds.

Classification

ORDER	Hymenoptera
FAMILY	Vespidae
SUBFAMILY	Vespinae
SPECIES	*Vespa crabro*
DISTRIBUTION	Europe and northern Asia
HABITAT	Woodland
NEST MATERIAL	Paper
DIET	Insects

ABOVE
UNWELCOME VISITORS
A small hornet's nest takes shape (and will continue to grow) in some roof beams, which are a good substitute habitat for hollow trees.

BEE HUNTERS

Hornets are tough predators. They are heavier than most other flying insects and so can easily out-compete their victims. They are a frequent predator of honey bees, and have a thick exoskeleton that protects them from stings. Studies show that hornets are especially sensitive to the alarm pheromones of honey bees, which the smaller insects produce to marshal defenders when attacked. Ironically, however, their alarm calls only serve to attract more predators.

CASE STUDY
Wood Wasp

This hymenopteran is also called a horntail due to the spike that protrudes from its abdomen. This is often mistaken for a sting but is actually a harmless ovipositor. The ovipositor is robust and is able to penetrate tree bark and deposit eggs inside a trunk—which can sow the seed of the tree's destruction. The wood wasp is distinct from its stinging cousins due to its lack of a narrow waist.

Classification

ORDER	Hymenoptera
FAMILY	Siricidae
SPECIES	*Sirex noctilio*
DISTRIBUTION	North Africa, Europe, and Asia
HABITAT	Pine forests
NEST MATERIAL	Fungus
DIET	Wood

DRILLING SOFTWOOD

The adult wood wasp is a big insect, reaching 1 1/2 inches (4 cm) in length. They need to be this size to wield enough force to cut through bark. They target mostly old and weakened softwood conifers, especially pines. After mating, the female drills a hole down to the layer of xylem vessels under the bark, where she lays two or three eggs and deposits a glob of mucus. This secretion contains the spores of a symbiotic fungus which will spread through the trunk ahead of the larva, keeping the wood damp and soft before it hatches.

TAKING TIME

The eggs hatch in around a week, but if the weather conditions outside the tree are poor, the larvae lie dormant inside the egg for up to several months, waiting for better times. When the temperature rises above 77°F (25°C), the larvae become active. If it gets too hot, above 86°F (30°C), then the larvae are likely to die inside the tree. At first, the larvae stay near to their mother's borehole, feasting on the fungus there. Once fortified, they mine their way through seams of xylem tissue following the spread of the fungus. This leads them into the heartwood. There, they are near to full size and take a turn so they are tunneling upward or down. When ready to pupate, the larvae then take another turn and head back toward the bark to make it easier for them to escape the tree as adults. When they are a couple of inches from the outside, the wood wasps pupate. The new adults then chew their way out, resulting in U-shaped tunnels running in and out of the tree.

INFESTATIONS

Male wood wasps develop from unfertilized eggs, and females lay ten male eggs for every female. This skewed ratio ensures that a female finds a mate among all those trees. She then lays up to five hundred eggs. When the population of wood wasps is high, the insects are forced to target healthy pines. The larval tunnels have a big impact on the tree. One common feature is flagging, where the branches on one side die back dramatically.

ABOVE
PUPAL STAGE
An immobile pupa of Sirex noctilio pauses to transform into an adult just below the surface of the tree.

RIGHT
EXIT HOLE
A newly emerged wasp has gnawed its way out of the tree in which it developed, leaving a neat hole.

1 First sting ←
Having dug a deep hole for its victim, the jewel wasp patrols the skies looking for a cockroach. When it finds a target, it delivers a double sting. The first is injected directly into the thoracic ganglion, a major junction in the cockroach's nervous system. This paralyzes the prey's legs.

Jewel Wasp Enslaves a Cockroach

Ampulex compressa, better known as the jewel wasp (or emerald cockroach wasp), is no shrinking violet. Instead of the yellow and black favored by so many of its kin, this tropical species from Africa and the Indo-Pacific has an emerald green body with a metallic sheen, and bright red thighs on its four rear-most legs. The jewel wasp also has an unusual power over its prey. Not only does it catch a cockroach much bigger than itself to feed its young, it force-marches it, still living, down into its own grave. The wasp does this by deploying a concoction of mind-control venom injected directly into the nervous system. Then, in a twist to the usual plot, the jewel wasp larva eats the zombie cockroach alive.

2 Second sting
With the cockroach immobilized, the wasp is free to deliver a second sting right into the victim's brain. The venom nullifies the cockroach's escape reflex and, as a result, even as the paralyzing effects of the first sting wear off, the cockroach will not make a break for it. Instead, it cleans itself before becoming sluggish in its movements.

3 Led to the grave ↓
The jewel wasp now chews off the cockroach's antennae, leaving about half attached, and drinks deeply on the hemolymph that trickles out. With a boost of energy, the wasp grabs one of the antenna stumps in its mouth and hauls the cockroach toward the pre-dug hole.

JEWEL WASP ENSLAVES A COCKROACH 101

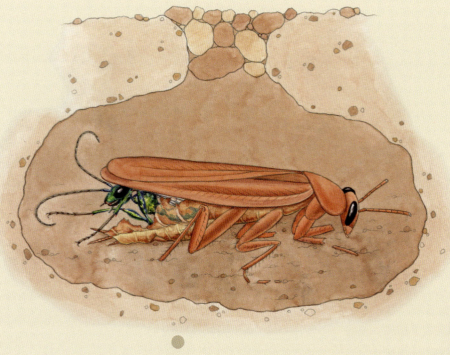

6 Becoming an adult ←
On day eight, the larva kills the cockroach, and pupates inside the dead insect. The rate of metamorphosis varies according to the temperature, but once emerged from the cocoon, the jewel wasp breaks out of the hollowed-out cockroach and digs its way to the surface. It lives aboveground for several months, mating several times, and leading dozens of cockroaches down to their doom.

5 Eaten alive
Three days later, a small wasp larva hatches from the egg. At first it consumes hemolymph from the cockroach's legs. After about five days of feeding like this, the larva then bites its way into the cockroach's abdomen and eats it from the inside out. As the larva prefers to eat the cockroach while it is still fresh and alive, it is careful to eat the vital organs last.

4 Buried alive ↙
The wasp lays an egg or two on the cockroach's legs and then seals up the death chamber. This is to stop parasites from getting in and stealing her prey. There is little chance that the zombie cockroach will revive and try to escape.

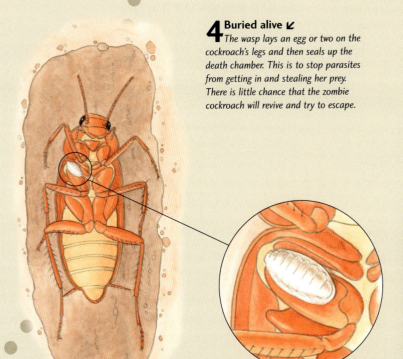

BELOW
PEBBLE SEAL
A jewel wasp collects a pebble to be used to seal off the entrance to its cockroach victim's death chamber.

CHAPTER FIVE
Bees

The honey bee, always busy, largely benign, and frequently helpful, is what we all think of when we hear the word "bee." We picture them in their hive, making honey, raising the young, and perhaps dancing as they tell each other where to find the next supply of pollen and nectar. However, as is a common theme in the world of insects, these social bees are just the tip of the iceberg. Along with the similarly social bumble bees (*Bombus* spp.) and the stingless bee tribe (Meliponini), honey bees (*Apis* spp.) make up less than 5 percent of the total number of bee species. In fact, most bees are solitary.

The twenty thousand or so species of bees occupy the taxonomic lineage, or clade, called Anthophila. They are all descended from a wasp that lived around 125 million years ago, in the Early Cretaceous. The fossil record shows that a novel lineage of plants, the earliest angiosperms, diversified around this time. Angiosperms can be found in abundance today. We know them as flowering plants, which reproduce through pollination. Many angiosperms, from grasses to oak trees, rely on the wind to spread pollen, but others employ the help of animals, especially insects, to carry out this job. As payment, the flowers dish out a sweet liquid meal of nectar. Some of the insects snaffle a portion of pollen to eat as well (there is plenty going spare.)

Essentially this is the feeding strategy of bees. Back in the Early Cretaceous, an ancestral wasp evolved out of its predatory ways to survive on what the flowers had to offer.

And today all members of the Anthophila, a name that means "flower lover," do the same. (Actually, it is nearly all; three bee species harvest carrion and feed a meaty liquid to their young, and there are even several wasps that, like bees, feed their brood pollen—this includes the pollen wasp, which is more closely related to paper wasps.)

FEATURES AND TRAITS

The common features of bees are best seen through a microscope. They include a feathery, or plumose, hairlike seta covering on the body. The antennae have an elbow-like joint, and comprise thirteen segments in males, and twelve in females. Other notable features can be found in their mouths, including lapping mouthparts for slurping up nectar and mandibles for biting, cutting, chewing, or manipulating wax or plant resins. Ultimately, the real distinction for bees is in the obligate consumption of pollen, nectar, and plant oils by the larvae.

In common with that of fellow hymenopterans, the bee life cycle involves mothers provisioning the young—with pollen, probably some nectar, and perhaps other flower products as well. Social species, such as honey bees, construct elaborate nests made from wax, where many dozens of developing young are fed on honey. However, there are many other kinds of solitary bee. They construct nests in different ways, from exploiting natural cavities in wood or rock to excavating burrows and building their own structures with

ABOVE
MOVING IN
A two-colored mason bee (Osmia bicolor) has converted an empty snail shell into a nest for its young.

ABOVE RIGHT
REDECORATING
A leaf-cutter bee hauls a piece of leaf for relining its nest in a crevice in dead wood.

RIGHT
OCCUPIED
The pupa of a red-tailed bumble bee (Bombus lapidarius) is resident in this subterranean nest.

resins and oils harvested from plants. They are named carpenters, masons, miners, and even plasterers accordingly.

There are also many bee species that do not work so hard and instead turn to more underhand methods to sustain themselves. These are cuckoo bees, also called kleptoparasitic, which means they "steal" the pollen stores of other bees. Much like the birds of the same name, cuckoo bees sneak one of their own eggs into the nest of a host bee. The cuckoo's egg hatches, feeds on the pollen meant for the host, and then emerges to repeat the cycle. In rarer cases, there are cuckoos who take over entire nests. Social robber bees raid entire colonies and take the resources back to their natal nest, while in socially parasitic bumble bees, a cuckoo queen assumes control over the workers of the host nest and imposes on them to rear her young larvae. Bees are not, therefore, quite as benign as we might imagine.

BLUEPRINTS
Honey Bee Comb

Honey bees form the genus Apis, which contains around seven to ten species. Like some wasps and all ants, they exist in a highly ordered society based around a single queen. They are eusocial, meaning there is reproductive division of labor: only the queen lays eggs. Honey bees live in anchored societies, where the sterile workers are specialized such that they cannot reproduce, and the queen's daughters help raise their sisters—and that means building a sturdy nest made from wax.

The colony's supply of honey is stored in the upper part of the comb. Foragers carry nectar back to the nest and regurgitate the dilute sugary liquid into the cells. Hive workers then fan the nectar with their wings, driving the water away to create a thicker honey. The sugar content is around 80 percent, so high that it works as a natural preservative. The cells store honey.

FIG. 1
BUILDING CELLS
A beehive emulates the natural home of a honey-bee colony. The box, woven basket, or whatever other structure forms the hive, must be dark like the hollow of a tree, a cave, or a mammal burrow, long abandoned. Inside, the bees construct an orderly comb of waxen cells. The beekeeper may have provided frames to maximize the colony size. In the wild, the combs are more haphazard tongue shapes, but still have the order conferred by the hexagonal shape of the cell.

A wax cap closes the cell during pupation.

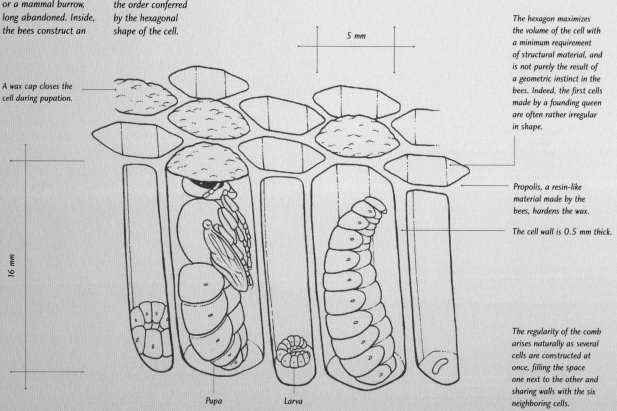

The hexagon maximizes the volume of the cell with a minimum requirement of structural material, and is not purely the result of a geometric instinct in the bees. Indeed, the first cells made by a founding queen are often rather irregular in shape.

Propolis, a resin-like material made by the bees, hardens the wax.

The cell wall is 0.5 mm thick.

The regularity of the comb arises naturally as several cells are constructed at once, filling the space one next to the other and sharing walls with the six neighboring cells.

FIG. 1 CELLS

HONEY BEE COMB

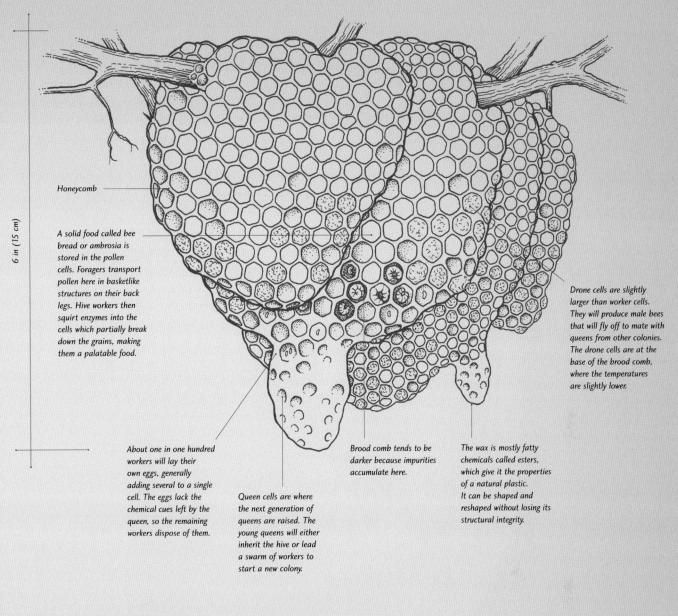

6 in (15 cm)

Honeycomb

A solid food called bee bread or ambrosia is stored in the pollen cells. Foragers transport pollen here in basketlike structures on their back legs. Hive workers then squirt enzymes into the cells which partially break down the grains, making them a palatable food.

About one in one hundred workers will lay their own eggs, generally adding several to a single cell. The eggs lack the chemical cues left by the queen, so the remaining workers dispose of them.

Queen cells are where the next generation of queens are raised. The young queens will either inherit the hive or lead a swarm of workers to start a new colony.

Brood comb tends to be darker because impurities accumulate here.

Drone cells are slightly larger than worker cells. They will produce male bees that will fly off to mate with queens from other colonies. The drone cells are at the base of the brood comb, where the temperatures are slightly lower.

The wax is mostly fatty chemicals called esters, which give it the properties of a natural plastic. It can be shaped and reshaped without losing its structural integrity.

FIG. 2
MAKING WAX
The building material of the honey bee (and its social cousins) is beeswax. Beeswax is made by the workers, who secrete it as flakes from glands on their abdomens. The raw material is honey, the colony's primary food. Honey is a mixture of sugars and water with some proteins and oils mixed in. The hive workers collect the clear, glassy flakes and chew them to mix in saliva and propolis. The latter is a glue made from resins and saps collected from plants and stored along with pollen supplies. The workers' handling of the wax also contaminates it with pollen, so it becomes steadily yellower as it is molded and carved into cells. The workers produce 2 pounds (1 kg) of beeswax from about 18 pounds (8 kg) of honey. However, that metric works better the other way round: A comb made with 2 pounds (1 kg) of wax (essentially from a million original flakes) can hold around 55 pounds (25 kg) of honey.

FIG. 2 COMB

Swarming

A honey-bee nest is built to last. While the paper nests of yellowjacket wasps are a one-year deal, the honey-fueled bee factory of social bees can run for years on end. Like other eusocial groupings, a bee colony can be regarded as a superorganism. A superorganism is a collection of organisms that work together so their group acts as a single entity. The honey bee superorganism reproduces by swarming. Primarily, a swarm occurs when the nest has grown too large to house and feed its workers. The resident queen recruits a cohort of workers who fly away together to start a new nest somewhere else. Young queens, newly hatched, will then battle to take control of the original nest once the primary queen has moved on.

1 The next generation ↓
The resident workers build large queen cells around the base of the brood comb. These special, protruding structures will be the nurseries for a new generation of bees. It will take sixteen days for the virgin queens to emerge once their eggs are laid inside. One queen will take over the nest. The others will probably be killed by their sister, or they may lead their swarm and start anew elsewhere.

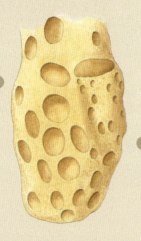

2 Preparing to leave →
About fourteen days before departure, the workers start their preparations to vacate the nest. They direct attention to raising the larvae of the new queen. They are fed exclusively on royal jelly, which is a protein-rich food secreted by the glands of the workers that nurse them. (Ordinary worker-bee larvae get a taste of this as well, but are also fed bee bread and honey.) The workers also stop feeding their current queen quite so much. During her resident phase, she's been too heavy to fly, and so needs to lose some weight before the swarming starts.

3 Food supplies
Around ten days out, the queen begins to rally supporters by piping. She produces short, high-pitched toots by flapping her wings at a very high rate. The workers may pipe in response and pass on the signal to others. They start to engorge themselves on honey and other foods in the nest. The aim here is to transport a food supply to the next nest in their stomachs.

4 Egg laying stops
Around a week before departure, the queen stops laying new eggs into the brood comb. The workers that stay in the nest will continue to raise the larvae in the cells. It takes around twenty-one days for a worker bee to develop from egg to adult. However, as the nest changes hands from old queen to new, there will inevitably be a short interim period when no new workers are emerging.

7 A new home
The bivouac will be in situ for a few hours at least, during which time scouts are sent out to find a more permanent home. The swarm will generally bivouac again near to suitable nest sites before the building of the first comb begins. The swarm will usually leave on a warm day, and needs at least two more days with similarly good conditions to secure a foothold in a new location. If the temperature drops when the bees are still out in the open, the future of the swarm is in peril.

8 Back home
The main swarm will generally leave before the virgin queens hatch. In large nests, a virgin queen could then recruit her own swarm that builds her a new nest while she takes her nuptial flight and mates with several drones. It is possible that the old nest will be left uninhabited by all these swarming queens, but usually a few young queens will attempt to take over, fighting it out until just one remains.

6 A temporary home ↑
Several thousand workers leave the nest. They coalesce around the marshaling point, and dance stop signals to others to indicate that this is where the swarm will gather. The swarm forms a living structure, a protective bivouac, around the queen. With the workers vibrating their wings, the bivouac is seldom quiet. Deep inside, the queen maintains the swarm with long episodes of piping, while the workers return to buzz running for short periods.

5 Time to go! ←
Scout bees have identified a branch or other structure a few feet from the nest. The queen then releases a chemical signal that triggers the swarm. The workers begin a behavior called buzz running, where they wiggle and jostle as they crowd on the comb and buzz their wings, occasionally making longer zigzag runs. This will continue for a few minutes as the swarmers prepare to leave.

RIGHT
ROYAL JELLY
Honey bee workers tend to a larval queen bee. The workers feed the larva exclusively on royal jelly, a secretion from glands in their throats.

MATERIALS AND FEATURES
Carpenter Bees

Carpenter bees (*Xylocopa* spp.) are a group of five hundred or so solitary species that chisel and carve nests in wood, usually dead—and that includes the timbers of houses. The nests are long tunnels, known as galleries, that can extend several feet inside a tree trunk. These galleries are used as a place of shelter in winter, but primarily serve as nurseries for the young.

The nests are always dug by a female in high summer. She scrapes her chewing mandibles on the hard wood and vibrates her body. This rhythmic motion creates a jackhammer effect that shifts the wood fibers more effectively. Nevertheless, it still takes the female several days to build her tunnel nest.

Using the grain

Typically the carpenter bee gallery is L-shaped. This is because it is easier for the bee to dig with the grain, grinding through the softer wood. However, often the female must make the entrance to the tunnel by digging across the grain. This means cutting through layers of darker and harder winter wood as well as soft paler wood that grows in summer. This is hard work, and it takes two days to dig down just 1 inch (2.5 cm). Then, once she reaches that depth, the female carpenter bee will take a 90-degree turn and dig with the grain.

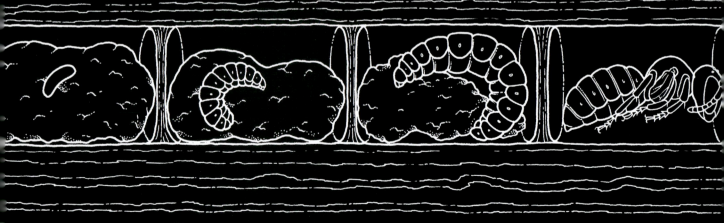

Finishing touches

The carpenter bee digs tunnels, or galleries, that are nearly perfectly circular. The tunnel is about $1/2$ in (1.5 cm) wide. Although the gallery might be branched inside it always has a single entrance, often with a pile of sawdust left by its builder. The bee uses its saliva to glue the sawdust spoil to build a papery wall that seals the tunnel.

Dividing walls

Once the nest is dug, the female converts it into a nursery. She places a mixture of pollen and nectar "bee bread" at the end of the tunnel and lays an egg on top. When the egg hatches, the bee larva eats the bee bread and grows into an adult inside its wooden cell. It takes about a day for the female to provision and seal one cell. Typically, she will build six cells then die. Her offspring emerges at around the same time, but not necessarily in the order they were laid. In some species, the young bees emerge in late summer but will return to this nest, or another suitable hole, to hibernate in winter. Generally the next generation stays put inside their mother's nest until spring, when they get busy building their own nest.

The young adults chew their way out, clambering over their less developed siblings and their dead mother as they head outside.

CASE STUDY
Tawny Mining Bee

It is perhaps no surprise that mining bees are known for building nests underground. The tawny mining bee, like all mining bees, lives in areas with well-drained sandy soil. A new generation of adults emerge from overwintering in subterranean nests in early spring and start to feed on a range of shrubs and trees. While the male will die soon after mating, the female's work is just beginning.

Classification

ORDER	Hymenoptera
FAMILY	Andrenidae
SPECIES	*Andrena fulva*
DISTRIBUTION	Western Europe
HABITAT	Grassland and woodland
NEST MATERIAL	Sandy soil
DIET	Nectar and pollen

UNDERGROUND NURSERY

Although the tawny mining bee can be found across western Europe, it is not an abundant species and sightings are uncommon. Females usually dig in sandy soils to create their nests, and several hundred of them can converge on a single patch of suitable ground. (Their structures might spoil the look of a well-kept lawn, but in fact help to keep the soil healthy.) Despite being crowded together, they display no social behaviors. The female breaks up the soil with her mouthparts and then sweeps out the spoil behind her with her hairy legs.

MAKING AN ENTRANCE

The nest opening connects to a vertical shaft up to 12 inches (30 cm) deep. The spoil from the excavation is arranged into a mound, known as a tumulus, that surrounds the entrance. The heap of soil acts as a dam to stop rain water trickling in.

BRANCHED BENEATH

The entrance shaft branches off in several directions, each terminating in a brood chamber. The female builder coats the walls of these chambers with secretions from glands on her abdomen. These waterproof the cell before she provisions it with bee bread, a mixture of pollen and nectar, which she hauls underground using pollen baskets on her thorax as well as her legs. She then lays an egg on the food supply. The egg hatches after a few days and will have pupated into an adult in less than a month. However, the weather aboveground is now too cold for the young mining bees to emerge, and so they lie dormant in their nurseries until spring.

TAWNY MINING BEE   111

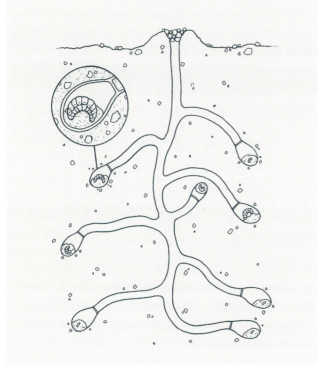

LEFT
IN THE MINE
The underground nest of the mining bee has several branches leading to a brood chamber for a larva.

ABOVE
HOME MAKER
A mining bee emerges from the nest. The insect is ½ in (1.5 cm) long, and her home plunges more than twenty times her body length below the surface.

CASE STUDY
Two-Colored Mason Bee

This species is one of the three hundred mason bees that make the genus *Osmia*. They earn this common name because they modify existing structures by adding mud, stones, and other materials. For example, many masons will find hollow nests such as those left by carpenter bees or wood-boring beetles and convert them into brooding cells by adding mud walls. The two-colored mason bee, however, builds a home for her young inside a hollow snail shell.

Classification

ORDER	Hymenoptera
FAMILY	Megachilidae
SPECIES	*Osmia bicolor*
DISTRIBUTION	Europe and western Asia
HABITAT	Grasslands and woodlands
NEST MATERIAL	A snail shell
DIET	Pollen and nectar

SHELL ACQUISITION
Male mason bees leave the nest several weeks before the females. Some emerge as early as February. They have been seen sheltering from the cold and wet inside their own snail shell. Their early rising ensures that they are ready to mate as soon as any females emerge in late spring. The now elderly male will die after mating, while the mated female begins her search for a shell. She is unfussy about the species of snail. The only requirement beyond being empty and largely intact is that she can lift it up. She will then fly with it to her preferred spot, generally near a good supply of food.

INTERNAL STRUCTURES
A decent-sized shell will provide space for five brood cells. These are located as far inside as the $1/2$-inch (12-mm) long bee can reach. A mass of pollen and nectar collected from wildflowers is packed into the shell and an egg laid on top. The mason bee then seals this cell with a wall of chewed-up leaves. Once the inner part of the shell has been fitted with cells, the mason seals up the open end of the whorl with a wall of sand grains, pebbles, and soil, to keep out predators and parasites.

CAMOUFLAGE
The shell has a natural camouflage, but the mason bee is not taking any chances. She will daube the outside with more leaf pulp to add more cryptic shading. Then she drops dozens of cropped blades of dried grass and twigs on top of the shell to break up its shape.

TWO-COLORED MASON BEE

LEFT
PACKED INSIDE
Breeding cells with a cache of bee bread can be seen packed into the central whorl of the shell, protected here behind a thick stone wall.

BELOW
FINISHING TOUCH
The bee prepares to drop a twig on a growing pile that covers, and hides, her shell nest.

CASE STUDY
Plasterer Bee

This Eurasian member of the genus *Colletes* has been given the name "plasterer" because of the way it seals the walls of its brood cells with a clear, natural plastic made from a mix of glandular secretions. This particular species is sometimes called the heather colletes, because it is seen harvesting foods from these plants in the heathlands of northern Europe. Elsewhere, it feeds on a wider range of wildflowers and shrubs.

NEST SITES

While some members of the family Colletidae seek out nesting sites to raise young inside natural spaces such as hollow plant stems or twigs, the plasterer bee is more typical in that it digs a burrow into loose soils, such as sandy cliffs or river banks. Often, several hundred bees will crowd together to set up home in the same area. The young males emerge from the nests early and swarm around the entrances, aiming to mate as soon as the females appear. Sometimes a female will be completely surrounded by a ball of males competing to mate with her.

WALLED OFF

A female only mates once and lays a single brood. She will dig a fresh burrow that has a central tunnel with branches off to several brood cells. The cell chambers are often wider than the tunnels. The mother will provision each cell with a liquid droplet of nectar and pollen (not the solid loaf of bee bread common with other bees.) So that this store of liquidy food does not soak through the walls, the cell must be waterproofed. Some plasterer bees glue their eggs to the ceiling of the cell, so the hatching lava may fall down into its food supply. Others float their eggs on the surface of the food droplet.

Classification

ORDER	Hymenoptera
FAMILY	Colletidae
SUBFAMILY	Colletinae
SPECIES	*Colletes succinctus*
DISTRIBUTION	Europe and western and central Asia
HABITAT	Heathland, moors, and sandy areas
NEST MATERIAL	Natural polyester and earth
DIET	Pollen and nectar

LEFT
LOOKING FOR A MATE
A male Colletes cunicularius, *or spring colletes, prepares to leave the nest in search of a mate.*

BELOW
SEALED OFF
Once each brood chamber (there are several in a nest) is provisioned with pollen and nectar, it is sealed with a layer of clear film and walled up with soil.

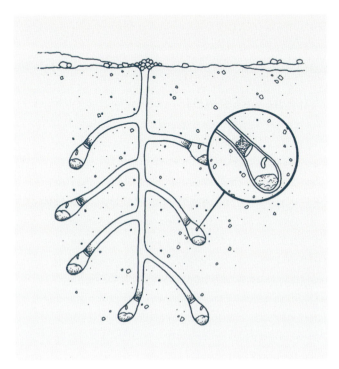

WATERPROOF COATING

The cell is lined with a natural polyester created by mixing secretions from the Dufour's gland in the bee's abdomen with chemicals in its saliva. When combined and spread on the soil, these materials dry into a clear film similar to a natural cellophane, which is why an even more prevalent common name for species of *Colletes* is "cellophane bee." The cell is sealed shut by the same material, making a transparent door across the narrower entrance tunnel. The access tunnel is then blocked with soil.

CASE STUDY
European Wool Carder Bee

Members of the genus *Anthidium* are known collectively as wool carder bees. Carding is a method of untangling the fibers of wool or cotton during textile processing using combing devices. Wool carder bees are so named because of the way they gather fluffy, wool-like plant fibers for their nests using comblike mouthparts. The European wool carder (*Anthidium manicatum*) builds nests for its larvae in whichever natural hollow in wood or rock is available.

INSIDE THE NEST

Once the female wool carder bee has identified a suitable cavity, she will start to fill it with soft bedding for her young. She creates a woolly ball that fills the space and is at least half her length. She then places a mass of pollen mixed with nectar. (Wool carder bees do not carry nectar supply in baskets on their legs, as is the norm for many bees, but attached to hairs on the underside of their abdomen.) The female then lays a single egg on the pollen and tucks it away under another, similarly sized ball of wool. This creates a secure space for the larva to hatch out and feed. The nest usually has room for a few of these cells, and the one nearest the entrance is sealed inside with a plug of earth and chewed leaves that are glued together with plant resins.

MALE AGGRESSION

Wool carder bees will mate several times a year and build multiple nests if conditions allow. While the females are busy working, the males are patrolling patches of flowers. They chase away rival males and species of bee that would otherwise deplete their treasure troves of food. Only female wool carder bees foraging for their broods are allowed to approach, and are quickly mated as they gather food. The males do not have a sting that delivers venom, but they nevertheless go through the motions of stabbing intruders with sharp spikes on their abdomens. Only the biggest and fittest males can control a flower garden, while the less well-endowed males will lurk nearby and sneak a mating when the opportunity presents itself.

Classification

ORDER	Hymenoptera
FAMILY	Megachilidae
SPECIES	Anthidium manicatum
DISTRIBUTION	Europe, Asia, and North Africa. Introduced to North America, South America, and most recently to New Zealand
HABITAT	Fields, meadows, and gardens
NEST MATERIAL	Plant fibers
DIET	Pollen and nectar

EUROPEAN WOOL CARDER BEE   117

LEFT
FURRY FIBERS
The fibers collected by the bees are trichomes, hairlike outgrowths that cover stems and leaves, often giving them a furry appearance.

BELOW
SOFT BEDDING
A bee larva is protected and insulated by the fluffy nest lining.

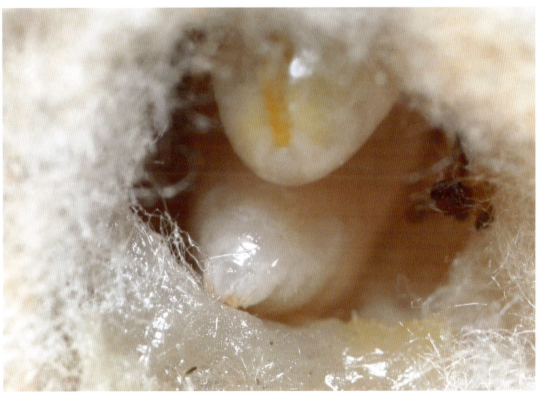

Leaf-Cutting Bee

Leaf-cutting bees (also referred to as leaf-cutters) belong to the genus *Megachile*. With about 1,500 species, this is the largest genus in the family Megachilidae, which it shares with mason, plasterer, and wool carder bees. As their common name clearly states, these bees cut out neat panels of leaf, leaving a distinct semicircular hole. They use leaf pieces to build cup-shaped cells for their young. Several of these are nested together in a tube-shaped cavity in a hollow plant stem or the abandoned home of another burrowing insect. If such habitats are not available, the bee will dig a tunnel into soft earth or chew out a space in rotting wood.

1 Reception party →
Adult female leaf-cutting bees come out from their nests in late spring. They are greeted by a crowd of males which have emerged a bit earlier. The young adults mate straight away, and the males will die soon after. The females, however, will live for several more weeks, during which time they will be very busy building a leaf-lined nest.

2 Tunnel vision
First, the mated female needs to locate a place for her nest. She may have access to suitable cavities left in wood or soil by other burrowing insects the season before. The bees will also inhabit hollow nesting structures in the "bee hotels" put out by gardeners. The cavity needs to be at least $1/3$ inch (7 mm) wide and can be 6 inches (15 cm) long. If no preexisting spaces are available, the female will likely dig one into the ground.

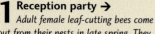

LEFT
LEAF SEAL
A female wood-carving leaf-cutter bee (Megachile ligniseca) heads into an insect hotel with a leaf to seal her nest.

LEAF-CUTTING BEE

3 Leaf cutting ↑
The name Megachile translates from Latin into "big lips." Insects lack lips, so this moniker refers to this genus's robust mandibles. These are the leaf-cutting tools. Making a series of bites, the bee will remove an oval section of leaf. Leaf-cutting bees are keen on using leaves of plants in the rose family, which includes apple trees, meadowsweets, and hawthorns. At most, the section will be around 1 inch (2 cm) long and 1/2 inch (1 cm) wide, but often they are half this size. The bee rolls up the leaf and holds it in its legs, carrying it as a scroll under its body as it flies to the nesting site.

4 Cell creation ↑
Starting in the deepest part of the nest, the bee folds the leaf pieces into a thimble- or cup-shaped structure. It takes four or five pieces to make such a structure, which then fits snuggly into the cavity. The female will next place a ball of pollen in the cup and lay an egg on top, creating a brood cell. She then caps the cell with several circular pieces of leaf.

5 Nested cells ↙
The next cell is constructed in the same way, fitting perfectly into the lip of the cell below. There is generally room for several cells, with some nests containing up to twenty of them. The construction phase is completed by late summer, upon which the adult female dies.

6 In the winter
Inside the nests, the larva will be ready to pupate into an adult form during the fall. However, development is stalled at this point, and the bee spends the winter in a dormant prepupal form. When spring returns, the bees undergo metamorphosis into adults. The adult male leaf-cutting bees transform first. They are smaller than the females and are reared in cells nearest to the entrance, leaving the slumbering females farther back. Their early exit ensures that the short-lived males maximize their mating opportunities.

CASE STUDY
Red-Tailed Bumble Bee

The red-tailed bumble bee is one of 250 species of bumble bee, most of which are eusocial and live in underground nests. Bumble bees are a common sight across the Americas, Europe, and Asia. These burly bees give out a deep, humming buzz, and they are one of the first bees to get in among spring flowers. Their dark body absorbs heat well, while the fuzzy coat of their hairlike setae allows them to retain such heat.

Classification

ORDER	Hymenoptera
FAMILY	Apidae
SPECIES	*Bombus lapidarius*
DISTRIBUTION	Western and central Europe
HABITAT	Meadows and woodlands
NEST MATERIAL	Wax
DIET	Pollen, nectar, and honey

FOUNDING A COLONY
Unlike the nests of their honey-bee cousins, a bumble bee colony does not survive the winter. Young queens fly out of the nest in fall and mate with the much smaller males that have left neighboring nests. Then the mated queen buries herself in the ground and becomes dormant for the winter. She will re-emerge in spring ready to build a nest. She may set up in a secluded grass thicket but usually moves into an abandoned rodent burrow or similar hole. She secretes wax from her abdomen, and uses this to build cup-shaped brood cells.

WORKERS EMERGE
The first clutch of daughters take over the work of building cells and provisioning them with pollen and nectar. The workers also build small honey pots where they can store food. If left here for a long period, the nectar will congeal into honey; bumble bees do not process the nectar in the same way as honey bees. Only a few grams are in store at any one time.

COLONY GROWTH
Over the summer the colony will number as many as five hundred. As the nest grows, the queen begins to lose control over her workers. Some may eat the eggs she lays and start to lay their own. The eggs laid by a worker are always males. In late summer, the queen lays a new generation of queens and her own clutch of males. They will soon fly away to mate, leaving their bumbling worker siblings to steadily succumb to the cold.

RED-TAILED BUMBLE BEE

LEFT
INSIDE THE NEST
The bumble bee nest is a haphazard mass of brood cells and honey pots.

CASE STUDY
Six-Banded Furrow Bee

This species belongs to a large family, Halictidae, collectively called the sweat bees. The reason for this name is somewhat prosaic. These little bees have the reputation of being attracted to perspiration, although nearly all bees will seek out this substance at times—for they are after the salts contained in the sweat. The six-banded furrow bee, named for the pale yellow stripes on its long abdomen, is larger than most, measuring about ½ inch (15 mm) long.

NEST STRUCTURE

Amazingly, six-banded furrow bees will live a solitary life in some places, sets up communal nests in others, and may even form facultatively eusocial organizations elsewhere. Their underground nests start with a tunnel that plunges diagonally down to a single chamber containing multiple brood cells sculpted from soil. The cells are lined with waterproofing secretions and stocked with a mixture of nectar and pollen. The bee targets flowers in the aster family, such as thistles and fleabane, for such food. One egg is laid in each cell, and it takes about five weeks for them to develop into adults.

THE NEXT GENERATION

The males in the nest's founding brood will fly off to mate. The females can either leave and dig a nest (though will probably have to hibernate first) or extend the access tunnel of their mother's nest and build their own brood chamber. The nest is now a communal society, where the females cooperate with building and maintenance, but provision only their own cells. They can also give up their reproductive drives, promote their mother to queen status, and devote their labor to raising their siblings.

Classification

ORDER	Hymenoptera
FAMILY	Halictidae
SPECIES	*Halictus sexcinctus*
DISTRIBUTION	Europe and western Asia
HABITAT	Areas with well-drained loamy soils
NEST MATERIAL	Soil
DIET	Pollen and nectar

SIX-BANDED FURROW BEE 123

LEFT
RESIDENT FEMALE
This sweat bee has built a nest in spring and is busy providing food for the brood cells inside.

BELOW LEFT
SWEAT BEE NEST
Each sweat bee nest has several brood cells reached by a tunnel. This illustration shows a nest inside a clay bank, seen from above and from the side.

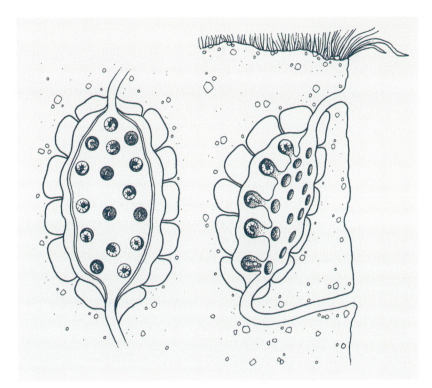

FLEXIBLE SYSTEM

Some of the females from each generation stay and some go it alone. The colony will only have two generations. The mated females in the latter generation return to the nest to hibernate until spring. In the north of its range, the six-banded furrow bee is more likely to live a solitary life, while in the south a form of eusociality is more common, with a single aggressive ruler. Communal nesting is found in between these areas. Experts believe that this species is not showing us how eusociality evolved, but reveals how the strict order of a bee colony can easily disappear as the bees return to a solitary life.

CASE STUDY
Stingless Bees

As the name suggests, this tribe of bees do not sting—but they will bite. Like honey bees and bumble bees, this tribe has corbiculae, or basketlike structures, on their back legs. Typically, and this applies to stingless bees too, these are used to carry pollen to the nest, where, like their better-known relatives, they manufacture honey. The five hundred or so species in the tribe Meliponini are eusocial, while a small set, robber bees, raid other stingless bee colonies.

BUILDING MATERIALS

As well as hauling pollen back to the nest, stingless bee foragers collect blobs of resin from a range of plants. This sticky, oozing liquid, released by trees, seals up any cracks and is filled with oils that smell bad to many insects. The stingless bee workers seek the materials out, blend them together, and mix them with their own wax secretions to make a pliable brown material called cerumen. (This term is also used for human ear wax, and the two materials look rather similar.)

NEST STRUCTURE

The bees build nests up trees, inside hollow cavities in the trunk, or under the ground. The nest always has only one entrance, often linked to a long tunnel that leads to the brood chamber. This is encased in an outer layer called the involucrum. (Further barriers are in place above and below to stop ants attacking via other routes into the cavity.) Inside, the individual brood cells are shaped like lozenges and stacked together in neat layers. Each layer stands above the next on pillars of cerumen. Aside from the brood comb are large irregular bulbs several times larger than any individual bee. These are storage pots for honey and pollen. While both are preserved in lesser quantities than by honey bees, there will be enough to sustain the colony year-round during periods of food scarcity.

Classification

ORDER	Hymenoptera
FAMILY	Apidae
TRIBE	Meliponini
DISTRIBUTION	Tropical and equatorial areas worldwide
HABITAT	Tropical forest and wooded savanna
NEST MATERIAL	A mixture of plant resin and wax
DIET	Honey, pollen, and nectar

STINGLESS BEES

LEFT
NEW EGG
The cell on the left has a freshly laid egg in a soup of pollen and nectar and will soon be sealed over.

ABOVE
HONEYCOMB OF BEES
*This brood comb of tropical stingless bees (*Melipona quadrifasciata*) from South America is surrounded by the larger conical honey pots.*

DEFENSE

Stingless bees have vestigial, or underdeveloped, stings, but the colony is not defenseless. With several thousand members, it possesses the weight of numbers, and many species will station soldier caste bees at the entrance to catch and kill any interlopers. At night, the entrance of a nest is sealed with a temporary lattice of cerumen containing a particular resin that is noxious to the bees' archenemies, the ants.

CHAPTER SIX
Ants

Ants are an insect success story. They are among the best known, most recognizable, and ubiquitous of insects. This is thanks chiefly to the fact that they are the most abundant insects on the planet. Sensible estimates suggest that there are 20 quadrillion ants at large in the world. Each ant weighs several micrograms, but their combined biomass exceeds all the world's wild mammals and wild birds added together. Their total weight of 12 million tons is equivalent to a fifth of the total biomass of humans. While it must be admitted that humans have had the most impact on the planet, ants surely come a close second.

Around 14,700 species of ants have been identified so far (with more to follow, no doubt) in the family Formicidae. The first ant fossils date back to around 120 million years ago, during the Cretaceous period. The most striking difference between ants, wasps, and bees in the order Hymenoptera is that ants generally lack wings. Only reproductive ants have wings for their brief nuptial flights. Additionally, ants have adopted more drab colors compared to their cousins, most being somewhere between black, red-brown, and pale tan.

All ant species use a eusocial system where a single queen rules a colony and produces all the young—mostly daughters. The queen's offspring become workers that find food, build and defend the nest, and raise their younger sisters. This social structure is underwritten by the unusual genetic relationships between generations. A human child inherits a set of genes from both parents, and so ends up with a double set. This is described as being diploid. Female ants (and wasps and bees) are also diploid. However, male hymenopterans are haploid, meaning they inherit only a single set of genes. This creates an unusual difference in the relatedness of siblings. Human siblings have a genetic relatedness of 0.5. In the way we inherit genes, that's a close link. Only identical twins have a closer one, and, crudely put, it underlines why siblings generally support each other. However, the mismatch in gene inheritance in ants means that ant sisters have a high relatedness score of 0.75. This is why a eusocial system works well for ants and their kin. Many nonbreeding females can fulfill the imperative to pass on their genes by supporting a few reproductive sisters. (The brother–sister relatedness in these insects is only 0.5, which works the other way.)

Nevertheless, not all ants are as strongly eusocial as others. The first ant species probably had small and simple colonies, where workers looked alike and were unspecialized for one task over another. Later in their natural history, especially after a rapid diversification of ants around 50 million years ago, species evolved much larger societies where ants were specialized into multiple castes. The size and anatomy of each caste reflects a particular job, be that defending the colony, foraging for foods outside, or caring for larvae.

ABOVE
FUNGUS GARDEN
A giant soldier leaf-cutter ant watches over her minute sisters as they tend to larvae surrounded by a crop of fungus, the species' food source.

LEFT
ANTHILL
This heap of pine needles and twigs is created and maintained by a colony of wood ants living beneath.

128 ANTS

BLUEPRINTS
Anthill

A typical ant nest is hidden from view, dug into soft, damp soil. Above the surface, all that can be seen is an anthill created from the material brought to the surface by the construction. The nest is founded by a single queen, after she has completed her mating flight. In cooler climes, this flight is likely to be part of a localized mass emergence of flying ants that takes place on warm days.

FIG. 1
AN ANTHILL
The widespread species Lasius niger, the common black ant, lives in underground nests across the temperate parts of all continents, barring Africa and Antarctica. The nests can take several years to grow, with the founding queen living for around fifteen years, cared for all that time by millions of workers that each live for only about two months. The early few clutches of workers are small and fast-developing, but workers grow steadily larger as the colony matures. The multiple spaces in the nest are used as living areas for the workers, food storage, and as brood chambers. Foragers tend to stay in the upper regions of the nest, nearer to the outside. This stops any infections they pick up spreading through the colony.

There are chambers in the anthill, mostly used by foragers.

The food the foragers provide (a mix of scavenged remains, nectar, and honeydew) is cleaned by workers before being taken deep inside, where the vulnerable larvae and queen are located. If there is enough food available, she will lay a new batch of eggs just as the previous clutch prepares to pupate.

The tunnels and chambers are dug out by the ants. They remove grains of soil using their pincerlike mandibles.

FIG. 1 ANTHILL

The flying males, or drones, die soon after mating, while the queens are decimated by predators as they search for a suitable nesting site. The lucky few will start a colony, dropping their now-unneeded wings and digging out a single chamber for a clutch of daughters. These offspring take over the work—and just in time. The queen is nearly starving by now, and will be sustained by generation after generation of daughters who feed her, enlarge the nest, and care for successive broods. The queen makes her presence felt over her workers by using chemical signals or pheromones. Her influence can only reach so far, however, and this limits the colony's maximum size. To go larger, a nest will need multiple queens ruling side by side. This is a strategy adopted by a minority of species.

FIG. 2
HEATING SYSTEM

The spoil dug out from the underground chambers is heaped up aboveground. This creates the familiar form of the anthill. There are several entry points from the hill into the nest below to ensure the ants can get in and out quickly. The hill also has ventilation shafts that allow excess heat generated from the several thousand ants underneath to escape.

The workers care for the young at every stage of development. The eggs are laid in designated chambers and once larvae hatch from the eggs, they are transferred to another chamber for feeding. The fully grown larvae then spin a cocoon. These pupae are moved to a separate location, and may be buried in soil to complete their development.

4 in (10 cm)

The side elevation of the anthill helps to catch more of the warmth from the sun at times of the day and year when it is low in the sky.

The solar energy of light hitting the side of the anthill warms the soil. An underground nest would not receive so much thermal energy from the sun.

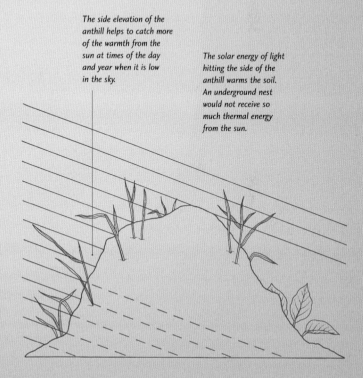

FIG. 2 OVERGROUND ADVANTAGES

MATERIALS AND FEATURES
Army Ant Bivouac

Army ant workers use their own bodies to create a nest for their queen and her larvae. These big ants are a feature of tropical forest habitats, mostly in South America. A queen driver ant, the *Dorylus* spp., of Africa, can be as long as a human thumb, but the soldiers are about 1 inch (2.5 cm) long, twice as big as the workers. The idea that this marching army devours everything unlucky enough to get in its way is overblown. However, smaller creatures, such as spiders and beetles, will scatter to escape the approaching killers. After all, a soldier ant has enormous mandibles for biting whatever threatens the colony, and for carrying heavy loads. Army ants do not form a distinct family but are an example of convergent evolution, and in Africa, Australia, and America such ants have all adapted to a nomadic lifestyle. This lifestyle is a response to the need to find a huge amount of food—half a million prey items a day—to supply the vast army, which can have millions of recruits. For that, the queen needs her workers to continually move to fresh feeding grounds.

A cross-section showing the interior parts of a bivouac.

Body building

The army ant nest is called a bivouac. It is constructed of the bodies of workers and soldiers, with each one gripping the back legs of the ant in front. This creates a raft of bodies that is built up in layers. Despite appearing chaotic, there is some order to the nest. The older, and more expendable, workers form the outer layers. Scans of the interior of the bivouac show that the thick mass of ants forming the outer region gives way to a less densely packed interior, where the queen and her brood are kept safe.

The bivouac

A bivouac is assembled on a tree trunk or hangs from branches. It may last for weeks or be broken down and moved each day. Typically, army ants follow a nomadic phase and then pause in a stationary one. The nomadic phase is synchronized with the development cycle of the larvae. It starts when the latest clutch of eggs—there can be more than a million—hatch. The larvae need food, so the bivouac dissembles, and the colony marches off and reassembles at a new location. Foraging parties collect food for several hours and then the army moves on. This daily march (or nightly in some army ant species) lasts for fifteen days, by which point the larvae are pupating and the demand for food reduces. The bivouac stays in one place for around twenty days (the queen lays after ten) and then the process begins again.

Marching

While on the move, the ants set a pheromone trail. The column following behind seeks out the strongest scent and so the individuals are constantly jostling each other toward the middle of a seething mass of ants. When the column is attacked or comes across an obstacle, the soldiers and workers will split from the column to investigate. These search parties can become detached and form "ant mills," where the ants are following their own scents around in a circle, and eventually die of exhaustion.

CASE STUDY
Carpenter Ants

The 1,500 or so carpenter ant species are named for the way they carve out nests in soft rotting woods. They do not eat the wood, as some termites do, but perform a similar function in decomposing dead wood. And that also means that these ants can be a damaging pest. A carpenter ant nest can seriously undermine the strength of timbers and make buildings unsafe. The black carpenter ant (*Camponotus pennsylvanicus*) is particularly destructive.

GALLERIES
A carpenter ant nest is a series of small chambers, or galleries, interconnected by narrow tunnels. The queen founds the nest by choosing a soft, rotten patch of wood that is easy to excavate with her mandibles. She targets fallen logs, or may start in a nesting hole or other entry point to a tree trunk left by another animal. The nest is swept clean of wood fibers, leaving a small pile of sawdust below the entrance. Once the colony is established, and workers take over the construction of the nest, they will extend it into harder wood, always following the grain.

HONEY POT
The Australian carpenter ant, *Camponotus inflatus*, lives in an arid environment, and so stores a honey-like food for times of scarcity. Instead of building containers for the honey, the sweet liquid is kept inside the body of workers, known as repletes. A replete's body is dwarfed by its highly swollen, honey-filled abdomen. Repletes make up nearly half of the total ant colony, and because they are too big to move, they must be fed by other workers. These same workers then drain the honey to feed themselves.

EXPLODING HEADS
There are some close relatives of the carpenter ants from Southeast Asia—until recently classified in the genus *Camponotus*—that have an unusual way of defending the nest. They form the genus *Colobopsis*, which is in the same tribe as carpenter ants, and its members nest in rotting wood in much

Classification

ORDER	Hymenoptera
FAMILY	Formicidae
SUBFAMILY	Formicinae
SPECIES	*Camponotus* spp.
DISTRIBUTION	Worldwide
HABITAT	Forests
NEST MATERIAL	Wood
DIET	Insects, carrion, nectar, and honeydew

CARPENTER ANTS 133

ABOVE
RIDDLED
As this rotten log falls apart it reveals the galleries formed inside by a long-gone colony of carpenter ants.

the same way. However, there is one big difference between the two. When a worker is in combat with a predator and cannot win, it contracts its stomach. This squeezes two enlarged glands in the head that are filled with poisons. The glands are squeezed so violently that the ant's head explodes, spattering its enemy with sticky toxins. This last-ditch attack may disable the attacker, or at least force it to retreat.

CASE STUDY
Asian Weaver Ant

The Asian weaver ant is one of two extant species that construct nests by gluing leaves together to build an enclosed space for the queen and her young. While its cousin, the common weaver ant (*Oecophylla longinoda*), lives in Central Africa, this species ranges through humid parts of India, Southeast Asia, and northern Australia. Weaver ants target trees that have wide, supple leaves. The queen starts by raising her daughters on the surface of a leaf.

Classification

ORDER	Hymenoptera
FAMILY	Formicidae
SUBFAMILY	Formicinae
SPECIES	Oecophylla smaragdina
DISTRIBUTION	Indo-Pacific
HABITAT	Tropical forests
NEST MATERIAL	Leaves and silk
DIET	Insects and honeydew

BRIDGING THE GAPS

The first generation of workers begins to build boxlike leaf shelters. As the colony grows, more nests are built nearby. A mature colony can have dozens of nests spread through the tree, housing hundreds of thousands of ants. Worker ants begin the process by pulling edges of neighboring leaves together, holding on with their feet and hauling the other edge closer with their mandibles. If the gap is too wide for one ant, several will climb into the space and create a bridge across. Once the edges have been closed up, they need to be glued in place. The adult workers cannot produce silk, but the larvae can. Other workers move along the join with a larva in their mouths and daub sticky strands that harden into silk sutures, holding the leaves in place. The simplest nests are a single leaf folded over, while more elaborate ones can be 20 inches (50 cm) across.

FIERCE DEFENDERS

Weaver ants often build nests in agriculturally important trees, such as mango, cashew, and cacao. Rather than exterminate them as pests from their orchards, cultivators see the ants as an ally, a biocontrol system that deters other pests. Weaver ants are predators that feed on other insects, including those that attack the fruits. The ants also tend to aphids and scale insects that supply secretions of sweet honeydew, and they will defend these flocks of bugs, driving away other ants. They have no sting but give a deep bite then squirt formic acid and other irritants into the wound, which amounts to the same thing. Fruit trees housing weaver ant nests produce a better harvest than those without, and there is no need for farmers to use pesticides on them.

LEFT
LEAF HOUSE
Weaver ants build in a large canopy and place the nest on an outward branch away from any threats on the busier trunk.

CASE STUDY
Florida Harvester Ant

There are several types of unrelated ants, mostly in the subfamily Myrmicinae, that have the common name "harvester ant." This is due to the way they collect seeds and sequester them underground in galleries for later consumption. The Florida harvester ant forages during the day and returns to large nests underground at night. The nests are made obvious at the surface by a wide but low mound of sandy soil with entrance holes near the middle.

UNDER THE GROUND

The nest of a Florida harvester ant can be as deep as 12 feet (3 m), but most of the living space and other chambers are clustered near the ground's surface. The nest has several helical or spiraled shafts that connect to labyrinthine lateral galleries or disk-shaped chambers. Most of the chambers are granaries, where foraging ants accumulate the colony's food. The chambers thin out as the shafts go deeper, and the queen and her brood are located in these safer areas of the colony. The youngest workers are employed in the deeper part of the nest, while the older workers are deployed as foragers and spend their remaining days nearer to the surface. This ant is unusual for a harvesting species in that it has two distinct castes of workers. Minor workers are less than $1/4$ inch (7 mm) long, while majors are almost twice the size. The latter are most commonly found on and near the surface and are distinctive for their seemingly outsized heads. These hold the large muscles required for their mouthparts to crack through hard kernels and protect the nest from attack.

HEAT TRAP

The purpose of the sandy mound above the harvester ant nest is not fully known. The workers clear this area of living plants and place tiny stones and dead leaves across it. One explanation for the mound's use is that it is a heat trap; and the stones have been placed there by the ants to absorb the sun's heat and transmit it to the nest below. This theory is supported by the way the ants often favor darker stones, especially charcoals, which absorb heat better.

Classification

ORDER	Hymenoptera
FAMILY	Formicidae
SUBFAMILY	Myrmicinae
SPECIES	*Pogonomyrmex badius*
DISTRIBUTION	East and southeast of United States
HABITAT	Dry, sandy areas
NEST MATERIAL	Soil
DIET	Seeds and insects

FLORIDA HARVESTER ANT

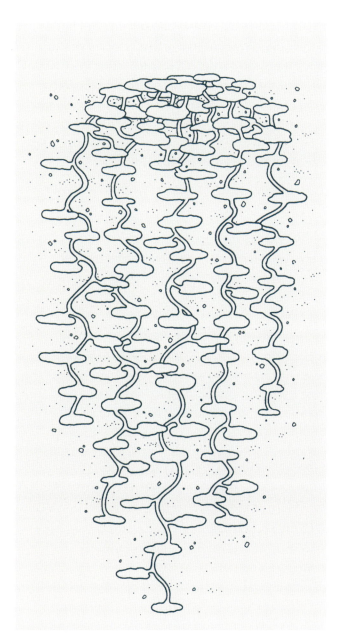

Florida harvester ants are also remarkable as a species in that they will generally move to a fresh nest once a year to avoid contamination risks. The new nest is dug a few feet away from the old one over the course of several days, with a steady stream of workers moving food and young to their new home.

ABOVE LEFT
DEEP STRUCTURE
A mature nest contains dozens of chambers connected by spiraling tunnels that penetrate several feet into the ground.

TOP
ENTRANCE CONE
A mound of sand marks out where a harvest ant nest lies beneath.

ABOVE
TEAMWORK
Harvester ants work together to place fragments on the sand of the entrance mound.

CASE STUDY
Acacia Ant

The acacia ant has a ready-made nest provided by the bullhorn acacia (or bullhorn wattle), a small tree that grows in low-lying humid areas of Central America. Most acacia plants fill their leaves with bitter-tasting chemicals that deter grazers. However, the bullhorn acacia recruits ants as security guards, and provides food and lodging in return. A big tree can have thirty thousand ants defending it with one of nature's most painful stings.

Classification

ORDER	Hymenoptera
FAMILY	Formicidae
SUBFAMILY	Pseudomyrmecinae
SPECIES	*Pseudomyrmex ferruginea*, *Pseudomyrmex spinicola*
DISTRIBUTION	Mexico and Central America
HABITAT	Bullhorn acacia plants
NEST MATERIAL	Acacia spines
DIET	Nectar, leaf nodules, and honeydew

MOVING IN

The bullhorn acacia is so named for the thick, pointed spines that grow among the leaves of its host tree. These spines are hollow, and a founding queen seeks out a large one from its scent. She bites a small hole into the spine and lays around twenty eggs inside. This first generation of workers help their mother to set up brood chambers in neighboring spines. The spines are about 1 inch (3 cm) long, which is ten times the length of the ant. They are already hollow, with a waterproof lining that stops the ant larvae inside from drying out. To begin with, the plant is not getting anything out of this new colony, but once the colony begins to number around five hundred members, it will put the ants to work.

DEFENSE FORCE

Once there are enough ants—which takes several months—the workers are able to devote more of their time to tending to the plant's needs. When an animal starts to eat the acacia leaves, the plant releases a pungent alarm odor. The ants rush toward the smell and sting the predator, which could be anything from a goat to a caterpillar. The sting causes a prolonged burning sensation, and one on the tongue is enough to deter even the most stubborn goat. Thanks to the ants waging a constant battle with any and (almost) all attackers, the acacia tree thrives.

LEFT
THORNY HOME
Acacia ants gather around their nest's entrance hole, gnawed into a hollow thorn.

FOOD SUPPLY

The acacia plant provides the resident ants with a nectar produced by extrafloral glands at the bases of the leaves. The tips of the leaves also develop nutritious nodules, named Beltian bodies, after their discoverer, Thomas Belt. The bodies are packed with oils and proteins, and are snipped off by worker ants and taken back to the brood chambers to feed larvae. The ants also supplement the food provided by their host by tending flocks of aphids and other plant bugs that exude a sweet liquid feces called honeydew. Ironically, this goes against the tacit agreement between ant and plant that the former keeps away animals that want to eat the latter, since the bugs can weaken the plant, and introduce disease.

Growing a Fungus Garden

The industrious leaf-cutting ants (*Atta* spp.) from the tropical and subtropical Americas are among the best known and most marveled-at ants. For one, they have huge colonies, each holding around 5 million individuals. The immense groups form highly ordered societies, in which the normal reproductive castes of queens and male drones are joined by four worker castes, each with a certain job to do. Secondly, these ants seem to have invented agriculture 10 million years before humans did! That is what all the aboveground activity is leading to: leaf-cutting ants are tending a garden underground by feeding a fungus with fragments of leaf. The ants don't eat the tough leaves, therefore, but instead feast on the more nutritious fungus.

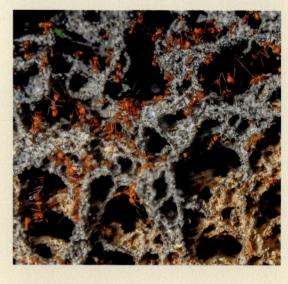

ABOVE
GARDENERS
Leaf-cutter ants tend to their fungus garden, formed of a mass of gray threadlike hyphae.

1 Searching for food ↘
The largest caste, known as major workers or soldiers, are out on patrol on the surface. They often feed on nectar and sap in trees rather than walking back to base. The soldiers are protecting the medias, which are the workers that search for sources of leaves. The medias are guided by the scent of the leaves which indicate that they contain the nutrients needed by the fungus garden. They seldom stray more than around 330 feet (100 m), and they leave a scent trail that leads other workers to any suitable trees they discover.

2 Cutting leaves
The workers climb into the tree and begin to cut out sections of leaves. They use their mandibles to cut pieces that are roughly their body length. It is estimated that a leaf-cutter colony consumes about 15 percent of the leaves in the area.

3 Back to the nest ↖
The foraging workers march back to the nest along the well-trodden trail, carrying the leaf aloft in their mouthparts. Often, smaller minor workers will be hitching a lift on the leaf piece. It is their job to protect the leaf from parasitoid wasps, which will try to lay eggs in it and so infest the nest.

4 Going underground
The leaf-cutter nest is a tall mound on the surface with several entranceways, plus smaller ventilation shafts. Once carried inside the nest, the leaf pieces are handed over to minor workers. They cut the pieces into smaller fragments and carefully clean each one, removing the outer waxy cuticle and any molds. If one source of leaves is found to be moldy, it is discarded and the foragers will search for a completely different leaf type to ensure the fungus garden stays healthy.

GROWING A FUNGUS GARDEN

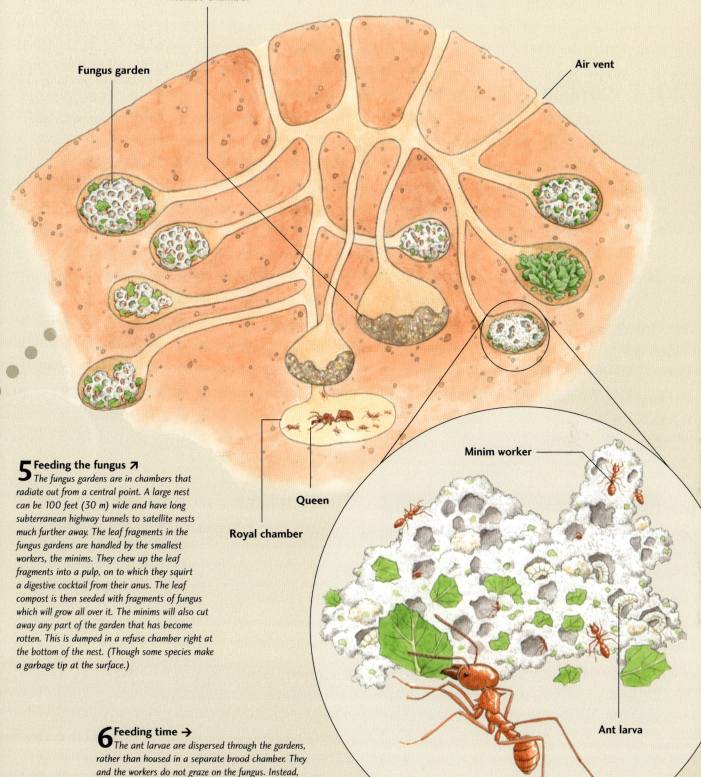

5 Feeding the fungus ↗
The fungus gardens are in chambers that radiate out from a central point. A large nest can be 100 feet (30 m) wide and have long subterranean highway tunnels to satellite nests much further away. The leaf fragments in the fungus gardens are handled by the smallest workers, the minims. They chew up the leaf fragments into a pulp, on to which they squirt a digestive cocktail from their anus. The leaf compost is then seeded with fragments of fungus which will grow all over it. The minims will also cut away any part of the garden that has become rotten. This is dumped in a refuse chamber right at the bottom of the nest. (Though some species make a garbage tip at the surface.)

6 Feeding time →
The ant larvae are dispersed through the gardens, rather than housed in a separate brood chamber. They and the workers do not graze on the fungus. Instead, the minims harvest swollen regions of the fungus called gongylidia. These are packed with sugars and other nutrients from the leaf, and any toxins and indigestible chemicals will have already been digested by the fungus.

CASE STUDY
Jet Black Ant

Jet black ants can be seen marching in narrow columns across the ground and in and out of trees as they forage for honeydew and insect prey. This innocuous species has quite the life story to tell. The shiny black ant builds large "carton" nests in the bases of old trees. The carton is a mixture of chewed wood pulp and saliva, which hardens into the animal equivalent of cardboard. It is unusual for ants to build with this stuff; termites use it a lot more.

Classification

ORDER	Hymenoptera
FAMILY	Formicidae
SUBFAMILY	Formicinae
SPECIES	*Lasius fuliginosus*
DISTRIBUTION	Europe and Asia
HABITAT	Woodland
NEST MATERIAL	Wood pulp
DIET	Honeydew and small insects

HYPERPARASITE
A queen jet black ant cannot start a nest by herself, so she flies into the nest of a close relative, kills the resident queen, and takes control over the orphaned workers. The usurper uses these workers to raise the first generation of true daughters. Soon, the original workers are replaced by invaders and take over the work. Mostly the queen targets host ants that build nests underground. One of those hosts is *Lasius umbratus*, the yellow lawn ant, whose queen has also already taken over the nest in the same way. That makes the jet black ant a parasite of a parasite—a hyperparasite.

IN THE NEST
Once the colony is established with a strong workforce, it relocates to a more suitable nesting site, such as a rotting oak, willow, or birch. The ants excavate the rotten heartwood, creating a cavity while constructing a labyrinth of chambers out of chewed wood. A subterranean extension is added to the nest with chambers lined with a rougher carton that has soil mixed in as well. This is a winter retreat for the colony.

BEETLE BURGLARS
Jet black ants are active night and day. At night, they are robbed by *Amphotis marginata* beetles that ambush them as they travel home along a well-worn trail. The ants are carrying food in their stomachs to be regurgitated for larvae and other nest mates. By copying the begging signals used by one ant to another, the beetle stimulates the passing ants to release their food supply directly into its mouth!

JET BLACK ANT 143

LEFT
TRUNK COLONY
Jet ants come and go from their nest hidden inside a rotten tree trunk.

LEFT
RESTRUCTURED
The dead wood inside a trunk has been sculpted by the ants into a labyrinth of chambers.

www.antstore.de

CASE STUDY
Argentine Ant

A native of the Paraná Valley that drains the borderlands between Brazil, Paraguay, Uruguay, and Argentina, the small, brown Argentine ant looks unremarkable. However, in just fifty years, this tiny creature—barely 2.5 mm in length—has spread to every continent and has become a pest on farms and in homes. The key to its success lies in its ability to build a nest almost anywhere, forming vast supercolonies made up of billions upon billions of ants.

Classification

ORDER	Hymenoptera
FAMILY	Formicidae
SUBFAMILY	Dolichoderinae
SPECIES	*Linepithema humile*
DISTRIBUTION	Introduced worldwide
HABITAT	Wetlands, forests, and domestic buildings
NEST MATERIAL	Bark, stone, leaves, and soil
DIET	Insects, spiders, and stored foods

EASY NESTERS

The Argentine ants' incredible success is due to their opportunistic nesting behavior, occupying any space that has a consistent level of moisture, but that never gets too wet. They do not need to modify these spaces much, beyond cleaning out fungus and other dirt. Once the space is full, the ants move on. In their native habitat, these ants are seldom found more than 6 miles (10 km) from the Paraná River, occupying the wetlands and a mix of other damp habitats. The water table is close to the surface here and so, to survive, Argentine ants have had to exploit a wide range of nest types. Mostly, they make shallow nests in the soil of leaf litter, but also under stones, behind the bark of tree trunks, and they sometimes set up home in old termite mounds. Translocated to urban and agricultural environments, the species is just as able to nest in cracks in concrete, in building cavities, and even in piles of garbage.

BUDDING

The nests of Argentine ants are generally small, with a few hundred workers. If there is room for more than this, then the nest will expand by adding more queens. The queens tolerate each other, and their offspring work together in the running of the nest as a whole. Argentine ant nests can reproduce by budding. This is where a queen gathers a group of workers from an established nest and sets off to start a new home. The ants in the new nest are treated as part of the same colony as those in the old one, and the workers will travel between nests without being attacked.

ARGENTINE ANT 145

LEFT
PIRANHA FEAST
Here, in their natural habitat, Argentine ants feast on the body of a washed-up piranha fish.

ABOVE
BUILD ANYWHERE
The success of Argentine ants comes from their ability to build a functioning colony almost anywhere, such as this crack in a rock.

SUPERCOLONIES

Reproducing in this way means that Argentine ants have been able to spread into vast supercolonies. The largest are found along the coast, with a big colony stretching up much of California. The world's biggest supercolony runs from northern Portugal to Italy via Spain and France. It has millions of queens and many billions of workers.

CHAPTER SEVEN
Termites

Termites are perhaps the preeminent insect architects. This reputation is mostly due to the fact that they often build above the ground—and they build big. The cathedral termites of Australia can construct mounds that are over 26 feet (8 m) tall. There are less well confirmed reports of chimneylike mounds made by *Macrotermes bellicosus*, a widespread Central African species, that are nearly 43 feet (13 m) tall! Beneath the ground, some ant species might be building bigger than this, but the termites still take the glory. Either of the examples just given easily qualifies as the tallest building constructed by creatures other than humans, and the structures also have a deep subterranean component. Termites do not all build on this scale, however. Some build carton nests (using a rough cardboard material) in trees. Others do not really build at all, but instead—as many a householder has discovered to their cost—tunnel through dead wood, including the timbers of buildings.

OPPOSITE
BUILDING BIG
The person sitting on top of this mound, built by termites in Brazil, illustrates its sheer scale.

RIGHT
VENTILATION
The top of this Kenyan termite mound has open-ended chimneys which allow the warm air from inside to escape.

ODD ONES OUT

Termites live in eusocial societies ruled by a single reproductive queen, but they are a completely different group of insects to ants, bees, and wasps of the order Hymenoptera that also live in this way. Instead, termites are more closely related to cockroaches, and share the order Blattodea with them. Termites occupy a subgrouping called the Isoptera, and that is again split into two groups, called the higher termites and lower termites.

Membership of these groups is based on feeding habits. Lower termites feed on wood and the insects employ a protozoan gut symbiont to digest the cellulose on their behalf. This is a rare example of a microbiome not based on bacteria. Roughly speaking, lower termites eat their way through dead or rotting wood, hollowing it out from the inside.

Higher termites, meanwhile, have a wider diet, many feeding on grasses and other plants and even animal feces. Several species, especially those that build the largest nests, still rely on wood as their source of nutrients, but will feed pulped wood to a fungus and eat that instead.

SOCIAL CASTES

Termite society differs from that of social wasps, bees, and ants in that it is ruled not just by a queen, but also by a king. These reproductive termites start out as winged adults that leave their natal nest. (The name Isoptera means "equal wings," reflecting the fact that their four wings are all similar in size.) After mating, the male does not die, but helps the queen to raise the first brood of young.

Termites are hemimetabolous, which means the young, or nymphs, resemble small, wingless versions of the adult. There is not, therefore, a helpless larval stage in which offspring demand constant care, as is the case with other social insects. As such, baby nymphs are put to work maintaining the colony almost from the get-go. The nymphs grow through a series of molts and can develop into different castes with a particular role to play. The workers are both male and female. The males become minor workers, and have small, nymphlike bodies their whole lives. Female workers can be elevated to major worker and then, at the final molt, some will transform into soldiers. This is the largest nonbreeding caste of all, tasked with defending the nest.

BLUEPRINTS
Termite Mound

If termites were scaled up to the size of a human, their respective mounds would each be far higher than today's "megatall" skyscrapers. Some mounds are not so high but have a diameter of several feet. They are formed from clays, saliva, and termite droppings—ingredients mixed by the workers and laboriously assembled, grain by grain,

**FIG. 1
INSIDE THE MOUND**

The Macrotermitinae are a subfamily of termites that grow their own fungus inside the mound. Members of this group produce arguably the most complex insect architecture of all. Inside, the upper part of the mound is an air cavity, known as the chimney, which may connect to the outside via a series of ventilation shafts. There is a smaller air cavity at the base of the mound, often below ground level, called the cellar. Above the cellar are interconnected chambers and cells which comprise the main living quarters for the termite workers. They retreat here when it is too hot—or too cold—outside. The largest space in this part of the nest is the royal chamber, where the queen and king produce eggs.

Above the living quarters is a fungus comb, where the workers tend a garden of fungus that grows on partially digested wood.

There are many entrances to the mound around the base.

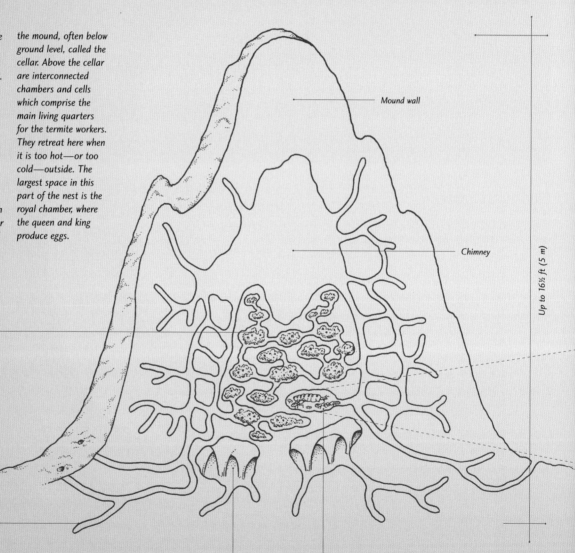

Side view · Mound wall · Chimney · Up to 16½ ft (5 m) · Cellar · Royal chamber

FIG. 1 THE MOUND

TERMITE MOUND   149

into tall edifices with thick walls to keep out predators, but also rain water and excess heat. There are a million or more workers active in a large mound, their bodymass weighing around 33 pounds (15 kg) in total. In one year, these insect architects are capable of shifting 550 pounds (250 kg) of soil. It can take decades to build a mound, and some are centuries old. The brutal simplicity of the mound's exterior is in stark contrast to the intricate structures within. As well as providing living chambers and a safe place to raise young, the mound is built with a ventilation system and, in many cases, a garden that provides food.

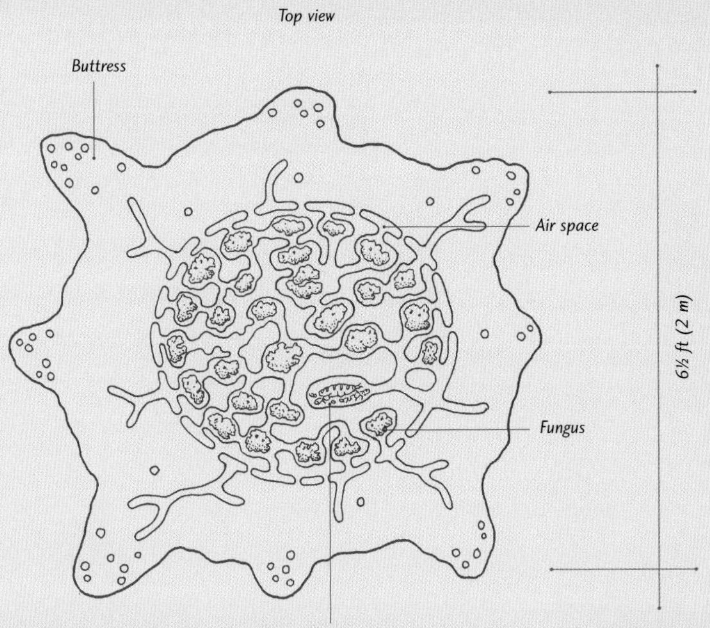

Top view

Buttress

Air space

6½ ft (2 m)

Fungus

Royal chamber

Eggs

The king is about 1 inch (2.5 cm) long.

A queen is well attended in her royal chamber, being fed by an army of workers and protected by a royal guard of soldiers.

The queen is much bigger than the king, as long as a human index finger and twice as wide.

FIG. 2
STRUCTURAL SUPPORTS

The mound is not completely round on the outside. There are numerous supporting buttresses around the mound's exterior to add to its structural integrity. The walls are porous so air and moisture can pass through them. However, they are prone to damage from heavy rains, which can wash away parts of the mound, so termites that live in areas with higher rainfall tend to use carton to build their nests. This is a material similar to cardboard, made from chewed wood pulp, fecal matter, and saliva. It is less sturdy than a clay mound but more resilient to rain.

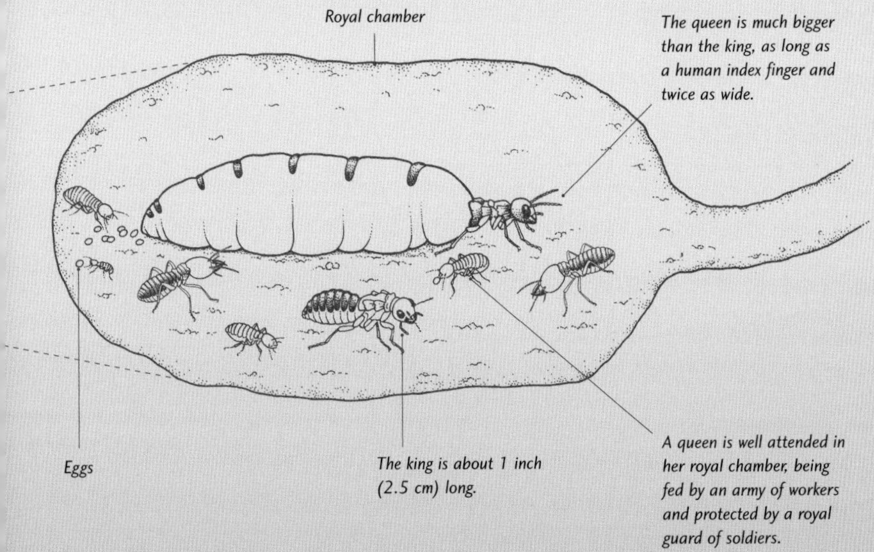

FIG. 3
ROYAL CHAMBER

Once the nest is established, the queen and her king stop work and concentrate on producing eggs. Both grow much larger than the workers. The king has the job of mating frequently as his queen produces eggs. The queen swells with eggs and in some fungus-growing species, she emits tens of thousands of eggs each day—a million every month. She might live for fifty years, every few years producing a brood of winged princes and princesses to fly away and begin a new nest. If one of the royal couple dies, then there are always immature workers in the colony that can develop into a reproductive form to take their place.

FIG. 2 & 3 SUPPORTS AND CHAMBERS

Gardening Fungus

Fungus-growing termites (Macrotermitinae) lack the symbiotic protozoa in their guts that help them to digest the nutrients from wood. Instead, they have entered into a partnership with a fungus called *Termitomyces*. The life cycle of the two symbionts is closely entwined. Indeed, the insects incorporate the fungus into the very fabric of their nests. The fungus garden, or comb, is located in the most humid region of the mound, so that the fungus will proliferate as quickly as possible. In a mature mound, the fungus fills up to 80 percent of the living space. This way of life is similar to that of leaf-cutting ants, and it is no surprise that those ants only live in the tropical parts of the Americas, while the Macrotermitinae live in the correspondingly tropical parts of Africa and Asia.

3 Feeding the fungus
The termites eat the wood and other plant foods but are unable to digest it. Instead, they use their droppings, which are filled with plant fibers and spores, to build the first parts of the fungus comb. The fungus begins to grow as diaphanous threads, known as hyphae. The hyphae combine to create a mycelium that is tended to by the termite workers.

4 Extending the comb
As the nest grows, the workers build more comb chambers from their wood-filled fecal pellets. The fungus in more established parts of the comb produces special spores which are moved around by the termites, eventually seeding more growth in new areas.

1 Breeding time ↓
Winged reproductive termites, known as alates, fly out of the mound and pair up with a mate. If all goes well, this queen and king will live together for many years. The mounds are also covered in termite mushrooms, the fruiting bodies of Termitomyces fungi, which spread spores in the wind. (The termite mushrooms are edible and include a West African species which is the largest edible mushroom on Earth. Its cap is 3 feet/1 m wide.)

2 Digging down ↑
The king and queen drop their wings and initiate the nest, digging down to build the first chambers. The spores from the mushrooms settle on the soil and food items in the area, and the fungus is carried into the nest by the termites. The king and queen work together to establish the fungus garden and raise the first workers.

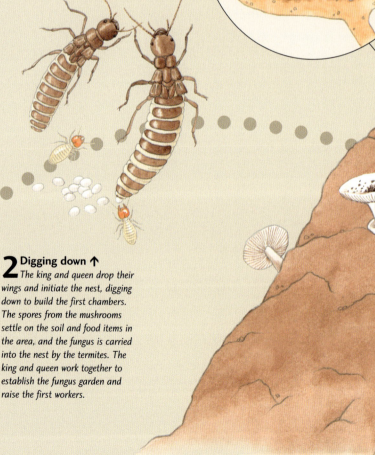

GARDENING FUNGUS   151

5 Food supply →
The workers that tend to the fungus comb are the only ones that eat the crop. They feed on the remnants of old combs but also harvest white nodules, called spherules or mycotêtes, that are produced by the living fungus. The termites then travel to other parts of the mound and feed the other insects with nutritious secretions from their salivary glands.

6 Mature mound ↓
The mound now hosts hundreds of thousands of termites, if not more. The king and queen will produce a new generation of reproductive sons and daughters. To spread spores, the fungus needs to grow long, stemlike structures up through the mound so mushrooms can sprout at the surface.

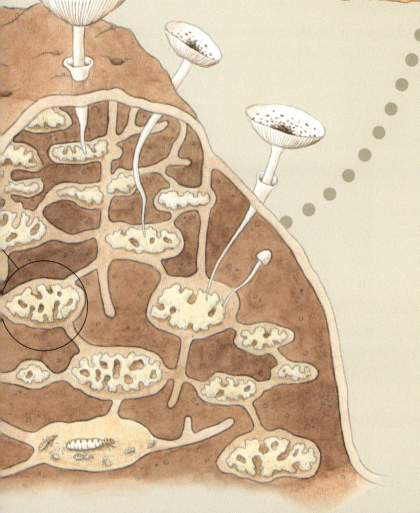

BELOW
TERMITE MUSHROOMS
The mushrooms that grow from a termite mound are mostly white with broad caps. There are currently fifty-two species of fungus known from Africa and Asia that are symbionts with termites.

MATERIALS AND FEATURES
Termite Ventilation

Like any large building, a termite mound needs an HVAC—or heating, ventilation, and air-conditioning—system. The termites inside require a steady flow of fresh air, a flow which brings in oxygen and flushes out the stale carbon dioxide. Additionally, the heat of all those millions of insects accumulates and must be released from the mound—but not too much on colder days. Likewise, the fungus combs perform best when maintained at a constant temperature and humidity. In structures built by humans, air pumps that circulate hot, cold, or dehumidified air are used to maintain temperate climes. Termites achieve the same results using a passive system that is powered by air currents and their own body heat, which has inspired human architects to rethink the way temperature is regulated inside buildings. And there are now numerous structures being built across the world where air-conditioning has been supplanted with passive cooling systems.

Central chimney
The simplest termite ventilation system is seen in mounds that comprise a chimney with a main ventilation shaft at the top. Here, the chimney top is tall enough to catch the breeze, as it whistles over the opening of the shaft. In something called the Venturi effect, the breeze pulls air out of the shaft. The warm air from the heart of the nest heated by the activity of the insects rises easily to replace the air lost through the shaft. In this way, a simple air current is set up, which is completed by fresh air being drawn in through lower side shafts that connect to the cellar beneath the fungus comb.

Cooling cellar
The air drawn into the mound is not necessarily cooler than the air inside. However, by being drawn down into the cellar, which is below ground level—often several feet down—the incoming air is cooled by the surrounding soil before it moves up into the main living area. The air in the cellar also flows around vanes of soil, where it picks up some moisture. This ensures the air in the nest stays moist and will not dry out the fungus.

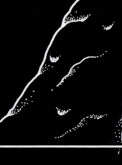

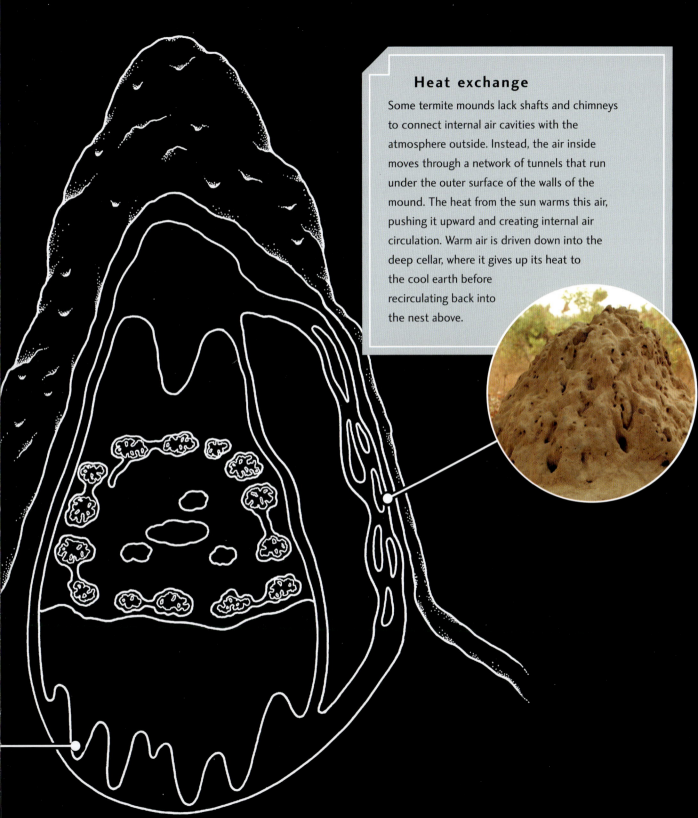

TERMITE VENTILATION

Heat exchange

Some termite mounds lack shafts and chimneys to connect internal air cavities with the atmosphere outside. Instead, the air inside moves through a network of tunnels that run under the outer surface of the walls of the mound. The heat from the sun warms this air, pushing it upward and creating internal air circulation. Warm air is driven down into the deep cellar, where it gives up its heat to the cool earth before recirculating back into the nest above.

CASE STUDY
Magnetic Termites

If one is ever lost in the grasslands of Australia's Top End, the northern region of the Northern Territory, it is possible to orientate yourself using the mounds of magnetic termites. The wider sides are aligned west–east, and the position of the sun can be used to determine compass directions. In the morning, the sun shines on the east-facing walls; in the afternoon, it will illuminate those that face west.

ALIGNED WITH THE SUN

Despite their common name, magnetic termites do not use Earth's magnetic field to align their nests. Instead, they build their homes in relation to the position of the rising and setting sun. The wider sides of the mound are orientated to be bathed in the warm light of the low sun as it rises and sets. That warmth is absorbed by the mound. However, when the sun is at its hottest and overhead at midday, it shines down only on the narrow surface at the top of the mound, so ensuring that the mound does not overheat.

SIZE AND LOCATION

Magnetic termite mounds can reach up to 13 feet (4 m) in height and about 8 feet (2.5 m) in width. A large mound can be around 3 feet (1 m) in diameter at the base but tapers to a jagged point at the top. The termites set up home in low-lying grasslands that are prone to flooding in the summer. When water rises to cover their surroundings, the termites stay inside the mound above the water level, surviving on a food store built up during drier times.

MAKING HAY

After the floodwaters recede, the grasses around the mound begin to die back. The termites then leave the nest and collect the dried grasses. Once cut into smaller pieces, the grass is piled up inside the mound, forming a supply of hay for the termites to eat. The termites have gut bacteria that help to digest this tough plant material.

Classification

ORDER	Blattodea
INFRAORDER	Isoptera
FAMILY	Termitidae
SPECIES	*Amitermes meridionalis*
DISTRIBUTION	Northern Australia
HABITAT	Grasslands
NEST MATERIAL	Soil and feces
DIET	Grasses

MAGNETIC TERMITES 155

ABOVE
ALIGNED
These tall, gravestone-like slabs of earth are invariably lined up so that the edges face north and south.

LEFT
AUSTRALIAN CLUSTER
Mounds are usually generously spaced from their neighbors, but in some places, such as here in the Northern Territory, Australia, the structures are clustered, creating what looks like a giants' cemetery.

CASE STUDY
Caatinga Termite

The Caatinga is an arid shrubland in eastern Brazil. This is an area dominated by thorny bushes which must withstand months of drought before a deluge of rain in midsummer. The wilderness appears largely empty of animal life, but it does exist—under the ground in the form of termites. Despite their presence going unrecognized for many years, these termites leave plenty of evidence of their existence—namely, around 200 million mounds of earth.

PILES OF SPOIL
The mounds made by Caatinga termites, known as murundus by local people, are not nests. Instead, they are piles of waste earth excavated by the termites to make tunnel systems under the ground. The insects climb to the surface through internal shafts and drop earth on top of the mound. Some of the mounds are four thousand years old, which means they were established around the time the Pyramids of Giza were being constructed. In the years since, it is estimated that the termites living under the Caatinga have shifted $2\frac{1}{2}$ cubic miles (10 cubic kilometers) of soil, which is the equivalent of building four thousand Great Pyramids.

LEAF HARVEST
The term Caatinga means "white forest" in the local Tupi language. This refers to how many of the trees and shrubs here are deciduous, and so stand pale and leafless during the long dry season. When the rains come, the white forest turns green within days. After a matter of weeks, the leaves start to shed again and fall to the ground. They constitute the only food source for the termites, and so the insects build tunnels that spread out around the mound leading to the surface. (It is this digging that adds spoil to the mounds.) Millions of termites race against each other to gather as many leaves as possible and bring them underground.

Classification

ORDER	Blattodea
INFRAORDER	Isoptera
FAMILY	Termitidae
SPECIES	*Syntermes dirus*
DISTRIBUTION	Eastern Brazil
HABITAT	Thorny woodland
NEST MATERIAL	Earth
DIET	Leaves

FAIRLY SPACED

The mounds are hard to see at ground level. Occasionally they might be tall enough to distinguish them from surrounding trees. From the air, the extent of the mounds is easier to see, and in fact some are visible on Google Earth. The mounds are roughly equally spaced. Termites from neighboring nests are tolerant of each other, but that tolerance fades with distance, and they will fight those from far-flung nests. The spacing of mounds is not due to colonies defending territory, therefore, but probably because the termites follow scent trails to find the nearest refuse point for their spoil.

ABOVE
BUILDING SITE
Each of the mounds is about 13 feet (4 m) tall and they are spread throughout an area larger than Great Britain!

CASE STUDY
Tree Termite

Not every termite species builds a nest under the ground or in an elaborate mound. As one might expect for insects that eat wood, termites also build nests in trees. Arboreal colonies are much smaller than the mound-building kind, with an upper limit of about 100,000 members. The colony forms wherever there is a source of damp, rotting wood, often around the base of an old tree. The main nest is then built later, higher up in the safety of the branches.

Classification

ORDER	Blattodea
INFRAORDER	Isoptera
FAMILY	Termitidae
SPECIES	*Nasutitermes walkeri*
DISTRIBUTION	Southeast Australia
HABITAT	Woodlands
NEST MATERIAL	Droppings, wood, and saliva
DIET	Wood

TREE NEST
Tree termites construct their nest from carton—a slurry of termite feces and saliva that is mixed with chewed-up wood. When wet, the carton is molded into sheets which become a natural form of cardboard. The outer layer of the nest is water- and wind-proof, but still quite fragile to the touch. The internal space is divided into chambers by papery walls.

SHELTER TUBES
The main food source for the termites is down around the base of the tree and in the root crown. Damage from decay and rot is more common here. A sizable portion of the termite colony is located in this area at any one time, and they travel back and forth to the main nest along shelter tubes. These are tunnels constructed of carton that hug the trunk. They are built under the cover of darkness and are seldom more than $1/2$ inch (1 cm) wide, but can run for several feet through the trees.

LETHAL COCKTAIL
This tree termite belongs to the subfamily Nasutitermitinae. All members of this group have soldiers equipped with an unusual weapon. Instead of a powerful bite, the soldiers have a fontanellar gun, a pointed nozzle positioned at the front of their heads. During combat, the soldier can squirt out from the gun a high-pressure spray of liquid glue, made from a cocktail of chemicals that combine just as they leave the nozzle. The force of the spray can kill attackers (normally ants) or will tangle their legs so they cannot move.

TREE TERMITE 159

ABOVE
HIGH LIFE
The main arboreal nest can be 66 feet (20 m) above the ground and is mostly built around the trunk where limbs are branching off.

CASE STUDY
Cubitermes Termites

Termite mounds are essentially built from a form of dried mud, and that material does not fare well in heavy rains. Species in the genus *Cubitermes*, found in the humid tropical zones of Africa, build aboveground nests in the shape of a mushroom or toadstool, or perhaps several of these standing one atop the other. The wide caps shelter the main body of the nest, ensuring that raindrops trickle to the ground without eroding the base of the mound.

Classification

ORDER	Blattodea
INFRAORDER	Isoptera
FAMILY	Termitidae
SPECIES	*Cubitermes* spp.
DISTRIBUTION	Africa
HABITAT	Forests and grasslands
NEST MATERIAL	Droppings
DIET	Humus and leaf litter

STRUCTURE

Cubitermes mounds are generally taller than they are wide. The central column is filled with chambers interconnected by a warren of tunnels, and the termites retreat to these chambers during periods of rain. The structure is built from soil and fecal pellets cemented together with saliva. These termites are soil feeders that eat nutrient-rich soils filled with humus and other detritus, and they have a pouch in their intestines that allows gut bacteria to ferment the tougher parts of this diet.

NEW RESIDENTS

An abandoned *Cubitermes* nest is not left to fall into ruin. Other termite species, known collectively as inquilines, will move in and do it up. Inquilines are a common feature in all nest-building termites, and often they have moved in while the original builders are still in residence. In these cases, the tenant species is eating a different food, such as the wood pulp in a fungus garden. There are non-termite species, most notably beetles, that find a home in mounds as well. These animals are called termitophiles.

OPEN CANOPY

Cubitermes nests are most common in woodland areas with an open canopy, where the trees' upper branches are generously spaced, allowing light and air to come through. Such locations are ideal for the nest because the widely spaced trees present less of a fire risk to the nest. The vegetation there also provides a good supply of nutrients, which the termites will then mix into the soil.

CUBITERMES TERMITES   161

**LEFT
MUSHROOM
MOUND**
*A distinctly shaped
Cubitermes termite
mound is pictured here
against the buttress
roots of a rainforest
tree in West Africa.*

CASE STUDY
Cathedral Termite

The cathedral termite is well-named. It builds awesome edifices that are intended to stand for the ages. The tall mounds, which are mostly over 16 feet (5 m) in height, sometimes almost double that, are constructed from a mud made from soil, fecal pellets, and saliva. This dries into a hard material that can be further reinforced by the addition of grasses. Construction lasts five years or more and the mound will stand for ten times that long.

LARGE HERBIVORE

In Western Australia, the cathedral termite mounds are more bulbous, whereas to the north they tend to be more spire-shaped. The termites spend a lot of time inside. They are prone to drying out in the arid conditions of northern Australia, and will only venture out when it is cool and humid enough (mostly at night). Cathedral termites are grass eaters. Their primary source of food is spinifex, or hummock grass, which is tough stuff. Silica crystals in the tips of the leaves make the plant unpleasant for larger animals to eat, yet the termites do not have this problem, and each colony consumes as much grass each year as a large cow will eat lusher grasses (if not more). In that respect, termites are the largest herbivores in this habitat.

SHIFTING DIET

The ancestors of cathedral termites arrived in Australia around 20 million years ago. It is supposed that these travelers came via rafts of vegetation that drifted from Indonesia or even all the way from South America and were from the Nasutitermitinae (as evidenced by the fontanellar gun of the soldier caste typical of the family). Members of this termite family are wood eaters, and they often build arboreal nests. However, evolution has meant that the incoming termites diverged to exploit grasses, thereby creating some of the largest mounds on Earth.

Classification

ORDER	Blattodea
INFRAORDER	Isoptera
FAMILY	Termitidae
SPECIES	*Nasutitermes triodiae*
DISTRIBUTION	Australia
HABITAT	Dry woodlands
NEST MATERIAL	Soil, feces, and grass
DIET	Grass

CATHEDRAL TERMITE   163

BREATHING MOUND

The vast cathedral mound uses its large surface area to drive ventilation inside the nest. By day, the sun warms the outer wall and the air circulating in channels near the surface. This air rises, and the cooler air further inside sinks, pushing fresh air into the base of the mound, where most of the termites are living. By night, the reverse happens. The outer wall cools faster than the interior, and so the flow of air through the mound changes direction.

ABOVE
MUD SPIRES
This mound in the Northern Territory shows from where the termites got their name. Sometimes the species creates simpler—but no less impressive— conical mounds.

Afterword
The Impact of Insect Architecture

As this book has shown, insects build dwellings, erect traps, excavate tunnel systems, set up refuse tips, and even tend to farms and gardens. And they have been doing it for millions of years—millions of years longer than we humans. The impact of all that building is obviously far-reaching. This is not only within the context of their own ecosystems—insect architects are a major factor in balancing biological activity—but also in how human architects and construction engineers can learn from their constructions. Insects have evolved the means to manufacture a range of building materials, from paper and wax to silk. What can we learn from these substances? Additionally, the structures of insect architecture and the processes by which they are built can be used—are being used—to improve human architecture.

WORKING FOR THE COMMUNITY

For every human on Earth, there are an estimated 1.4 billion insects. Termites, ants, and social wasps and bees make up a large proportion of this number, but most will belong to insect orders that do not build at all. Nevertheless, this figure demonstrates the cumulative potential of insects to impact the world around them. While a single carpenter bee, case-making caddisfly, or chimney-building cicada has little impact on its own, in their billions they are a force that changes the complexion of the natural world—and the human-made world, too.

As part of the broader ecosystem, insects are helping to maintain a stable natural environment. To illustrate one of many examples, predatory wasps are crucial for controlling the populations of other invertebrates, such as caterpillars, which are a popular food item used to provision brood chambers and nests. And in so doing, the wasps are protecting tree and plant populations that might otherwise be stripped by the caterpillars. Similarly, beetles of various kinds are important detritivores, or the consumers of waste and dead material. Without them, the natural world would be covered in feces and dead bodies!

WEBS OF NOURISHMENT

This kind of activity is all part of the food webs that transfer energy through an ecosystem and circulate the nutrients that life needs. Insects that build in soil, not least leaf-cutting ants and termites, help with this in another way. Their subterranean earthworks aerate the soil, which boosts root growth and helps the plants aboveground—on which the insects rely as food—to thrive.

Of course, insects are themselves a valuable source of food for other animals. In South America, the giant anteater uses its burly forelegs equipped with hooked claws to dig into termite mounds. Using its 2-foot (60-cm) tongue, which flicks in and out of its tiny mouth on the tip of its long face, an anteater can consume 35,000 termites in a single meal (and eat surprisingly few ants, despite the name). Its tongue is coated with sticky saliva and has tiny hooks that grab hold of several termites at once. In Africa and Asia, this niche is filled by pangolins, which look even more outlandish with their horny scales armoring their bodies, and the aardvark, with its piglike snout.

In Australia, the ant-eater role is taken on by echidnas, some of the few egg-laying mammals still living. Also among the spinifex hummocks in northern Australia, ants are targeted by the thorny devil, a slow-motion ant hunter. Despite being about the length of a hand, the persistent lizard can scoff three thousand ants a day.

BELOW
BABY FOOD
A potter wasp (Eumenes coarctatus) has paralyzed a captured caterpillar and prepares to stuff it into its pot-like nest for the larva to eat as it grows inside.

WHAT'S IN IT FOR US?

The benefits that humans receive from a healthy natural world come under the umbrella term of "ecosystem services." These include such perks as beauty and recreation as well as heavier topics such as control of climate and flooding and the mitigation of pollution. However, the really significant benefits of a thriving natural world are water and food, and many insect architects, most notably bees, have a large role to play in the production of humans' food supply.

It is estimated that two-thirds of crop plants rely on bees for pollination. The list is too long to even summarize effectively. Bees have built on the supply of flower nectar and pollen to create their elaborate nests and complex societies. In return, the flowers get to reproduce and grow into the fruit and vegetables we eat. That is one good reason among countless others that insect architects should be understood and valued.

INSPIRING ARCHITECTURE

Insect builders can also show human architects how to build better. For example, the Eastgate Centre, a commercial building in Harare, Zimbabwe, has a shaded cellar and atrium connected to dozens of funnel chimneys, inspired by the structures of termite mounds. Cool air from shaded areas is drawn into the base of the funnels and upward by warmer air from the upper stories as that rises up the chimneys and is vented out the top. This design also takes inspiration from other biology, such as cacti. It uses 50 percent less energy than a typical building and is generally 5°F (-15°C) cooler inside than out.

This form of biomimicry is also being extended to the construction materials that were invented by insects. A composite is material that combines the properties of several components. To a materials scientist, a mixture of dried soil and plant material cemented together with a liquid glue and then sealed in a waterproof layer is an example of a composite material. The hexagonal comb arrangement that gives honey bee hives their structural integrity is used widely in human construction and technology. For example, the wings of today's passenger jets are made from composites that are just as strong and flexible as metals but lighter in weight thanks, in part, to using honeycomb sections in the interior.

The other great breakthrough in insect engineering is silk, although, admittedly, spiders make this stuff as well, and probably do it better. Spider silk, or something like it, has been heralded as a new super-substance that could be used to make everything from space elevators to artificial muscles. A breed of goats has been genetically engineered to produce silk (from spider genes) in its milk. These goats make silk much faster than the spiders can manage themselves, so researchers have plenty of it to study.

A final insect contribution to building is an aesthetic one. "Swarm architecture" uses the same iterative approaches deployed by insects, where each new phase of construction is a larger version of the one that came before. This results in spiral complexes of several buildings of varying sizes that is a more pleasing use of space than more traditional architecture, or single buildings that are structurally sound but have more efficient yet amorphous shapes.

Insect constructions were here long before us, and perhaps as we humans build a sustainable future, insect architects will lend us a guiding hand.

OPPOSITE
HUMAN SCALE
The Richard Gilder Center for Science and Education, at the American Museum of Natural History in New York, is an example of architecture inspired by natural structures such as the galleries and chambers inside a termite mound (opposite bottom).

Glossary

Abdomen: Rear section of an insect's body containing the digestive and excretory organs and the genitalia (which includes the sting).

Alga: Simple, generally microscopic aquatic organism that photosynthesizes like a plant but lacks roots, stems, or leaves.

Algorithm: Step-by-step instructions designed to perform a specific task.

Anamorphic: Form of development in hexapods characterized by the addition of abdominal body segments at each molt until adulthood. The immatures and adults otherwise appear similar.

Angiosperm: Group that contains plants that produce flowers and seeds enclosed within a fruit.

Antenna: Sensory appendage found on the heads of insects, often used to detect odors and other chemical signals.

Asexual: Form of reproduction that only requires one parent (the female) and does not involve the fusion of sex cells from two parents.

Bioluminescent: Organism that produces light through chemical reactions within its body.

Carnivorous: Organisms that hunt prey, feeding primarily on the bodies of other animals.

Carrion: Decaying flesh of dead animals.

Cercus: One of a pair of appendages (plural cerci) on the rear-most segment of many arthropods, including insects, such as earwigs. The cerci serve a sensory function and may also be used in mating or defense.

Cocoon: Protective case spun by the larvae of certain insects in which to pupate.

Convergent evolution: Process by which unrelated organisms independently evolve similar traits by adapting to similar environments.

Cremaster: Hooklike structure at the end of a butterfly or moth pupa that is used to attach the sedentary cocoon securely to a surface.

Diatom: Single-celled algae encased in intricate silica tests, or cases.

Eusociality: Most complex level of social organization in animals in which species exhibit cooperative care of offspring, a reproductive division of labor with castes (queen, worker), and overlapping generations. All ants and termites, and some bees and wasps, are eusocial, as are a few other kinds of insects.

Exoskeleton: Hard outer structure that supports and protects the body of an arthropod, including the insects.

Family: In biological classification, a rank between order and genus.

Fungus: Kingdom of spore-producing saprophytic organisms that grow over and through foods, digesting them externally.

Gall: Plant growth induced by parasites such as insects, fungi, and viruses. Galls often house the organisms that stimulate their production.

Genus: A rank in biological classification above species and below family that contains a group of closely related species (plural genera).

Gongylidia: Specialized nutrient-rich structures produced by certain fungi as food for animal symbionts, such as leaf-cutter ants.

Hemimetabolous: Insect development characterized by incomplete metamorphosis, where the young form, nymphs, resemble the adult form and gradually mature without larval and pupal stages.

GLOSSARY

Hemolymph: Circulatory fluid in insects and other invertebrates, which functions similarly to blood by transporting nutrients and waste, but it fills the body cavity and is not confined within blood vessels.

Herbivorous: Organisms that feed exclusively on plants.

Hexapoda: Class of arthropods in biological classification and including Entognatha and Insecta, characterized by, among other features, six legs.

Holometabolous: Insects that undergo complete metamorphosis with distinct life stages: egg, larva, pupa, and finally adult (or imago).

Hyperparasitoid: Parasite whose host is also a parasite.

Inquiline: Animal that lives within the nest or structure of another species, mostly harmless to the host.

Instar: Developmental stage between molts in insects and other arthropods.

Kleptoparasitic: Organisms that steal food or nest space from other species.

Labial: Relating to the labium, an insect mouthpart, serving as a posterior or lower boundary to the oral cavity.

Larva: Juvenile form of an animal, especially that of certain insects, that looks distinctly different from its adult stage.

Malpighian tubules: Excretory organs in insects and some arachnids.

Mandible: Insect mouthpart used for grasping, crushing, or cutting food.

Metamorphosis: Transformation process in which an insect (or other animal) develops from an immature form to an adult.

Micron: Unit of length equal to one-millionth of a meter (1 μm).

Midden: Refuse heap created by animals (or prehistoric humans).

Mycangia: Specialized structures in some insects for transporting fungal spores.

Nymph: Immature form of hemimetabolous insects, those that undergo incomplete metamorphosis. Nymphs resemble smaller versions of adults but without fully developed wings or reproductive organs.

Order: A rank in biological classification above family and below class.

Paraneoptera: Major lineage of Pterygota that includes bark lice, true lice (parasitic), thrips, whiteflies, planthoppers, true bugs, and their kin.

Parasite: Organism that lives on or in a host, deriving nutrients at the host's expense.

Parasitoid: Organism, most often a wasp, whose larvae develop inside a host organism, ultimately killing it.

Parthenogenesis: Type of asexual reproduction in which development of embryos occurs without fertilization.

Pheromone: Chemical signals released by an organism to attract mates.

Polyneoptera: Major lineage of Pterygota that includes a diverse assemblage of insects, such as stoneflies, earwigs, stick insects, web spinners, ice crawlers, grasshoppers, crickets, katydids, roaches, termites, and mantises.

Pronotum: Upper surface of the first segment of an insect's thorax. This is often a prominent plate behind the head that is used for protection or ornamentation.

Pseudopod: Fleshy projections on a larval body that function like limbs or feet.

Pterygota: Major lineage of insects, characterized by the presence of wings. All insects are pterygotes with the exception of two groups (the primitively wingless Archaeognatha and Zygentoma).

Pupa: A life stage in holometabolous insects between larva and adult; the organism at this stage is generally inactive.

Salivary: Relating to saliva or the glands that produce it.

Species: Most basic unit of biological classification. A species is a group of organisms that can interbreed and produce fertile offspring, sharing common characteristics.

Spinneret: Organs through which some insects (and all spiders) secrete silk.

Thorax: Middle section of an insect's body, to which legs and wings (when present) are attached.

Triungulin: First larval stage of certain beetles that is characterized by being highly active and having three claws.

Venomous: Organisms that produce venom, a toxic substance that is actively delivered via bites or stings to a target for defense or capturing prey.

Xylem: Plant vascular tissue that transports water and minerals from the roots to the rest of the plant.

Index

Page numbers in **bold** type refer to pages containing illustrations.

A
Acanthopsyche atra 49
ambrosia beetle gallery **20–1**
Amitermes meridionalis 154
Amphotis marginata 142
Ampulex compressa **100–1**
Andrena fulva **110–11**
Anthidium manicatum 116–17
anteaters 164
anthills **127**, **128–9**
antlions 64, **74–5**
ants 42, 126–45
    acacia **138–9**
    Argentine **144–5**
    army ant bivouac **130–1**
    Asian weaver **134–5**
    carpenter **132–3**
    case studies 132–9, 142–5
    Florida harvester **136–7**
    fossils 126
    fungus garden **127**, **140–1**
    jet black ant **142–3**
    materials and features 130–1
    marching **131**
    mills **131**
    weaver **12**
    yellow lawn 142
aphids 19, 78, 80, 139
Apis spp. 102
Arachnocampa luminosa **72–3**
Archaeognatha 8
Archipsocus nomas **58**, **59**
Archips rosana **50–1**
Arthropoda 10
Atta spp. **140–1**
Austroplatypus incompertus 20

B
bagworms **48–9**
bark
    beetles **17**, 20
    lice 40, 43, **58**, **59**
    booklice 58
bees 42, 102–25, 166
    bumble 102, **103**
    carpenter **108–9**
    digger 32, **33**
    European wool carder **116–17**
    features and traits 102–3
    honey 97, 102
    honey bee comb **104–5**
    leaf-cutter **103**, **118–19**
    mason **103**
    plasterer **114–15**
    red-tailed bumble **120–1**
    six-banded furrow **122–3**
    stingless **124–5**
    swarming **106–7**
    sweat **122–3**
    tawny mining **110–11**
    workers **14**, 120
beehives 10, **14**, **104**
beetles 42
    ambrosia 20–1
    bark **17**, 20
    blister **32–3**
    carrion **22–3**
    dor **25**
    dung 15, **17**, **24–5**
    longhorn 26, 27
    spruce bark 17
    tiger 28, 29
    two-colored mason 112–13
beetles and bugs 16–39
    case studies 26–39
    materials and features 20–5

    wax and foam **18–19**
Belt, Thomas 139
Beltian bodies 139
bite 13, **74**, **75**, 76, 124, 134
bivouac, army ant **130–1**
Blattodea 9, **9**, 147, 154–63
Bombus lapidarius **103**, **120–1**
Bombus spp. 102
Bombyx mori 40, **41**, **44–5**
bristletails 8
brood cells
    ants **128**, 138, 139, **141**
    bees **14**
    common wasps **86**, **87**
    honey bees **105**, **106**
    hornets **15**
    hover wasps 94
    leaf-cutting bees **118**
    plasterer bees 114, **115**
    potter wasps 92, **95**
    red-tailed bumble bees 120, **121**
    six-banded furrow bees 122, **123**
    stingless bees 124, **125**
    tawny mining bees 110, **111**
    two-colored mason bees 112, **113**
buffalo gnats 67
bugs 9, **9**, 16–39
burrows 28, **29**, 36, **37**

C
caddisflies **9**, 42, 64, **66**, 67
    cases 68–9
    net-making **70–1**
camouflage 48, 80, 81, 112
Camponotus inflatus 132
Camponotus pennsylvanicus 132
carrion beetles **22–3**
cerci **34**, 38
Cercopoidea 18, **19**

INDEX

chimneys **146**, **148**, **152**, 164, 166
Chrysopa perla **78–9**
Chrysoperla carnea **80–1**
cicadas 17, 19, **30–1**
Coccoidea 18
cockroaches **100–1**, 147
cocoons
  ants 129
  antlions **75**
  common green lacewing 80
  jewel wasp **101**
  moths **7**, 40, **41**, 43, 44, **45**
  pollen wasp 90
Coleoptera **9**, 16, 26, 28, 32, 42
Colletes succinctus **114–15**
Colobopsis 132
communal nests 42, 43, 52, 54, **55**, 56, 122, 123
cricket orchestra 36
crickets 8, 16, **36**, 37
cryophiles 38–9
Cubitermes spp. **160–1**
Curculionidae 20
Cynipidae 88–9

D
damselflies 8
dance 54, **107**
defense 32, **58**, 64, 124, 138
Dermaptera **8**, 16, 34
diets
  algae 58
  ants 28, 76
  aphids 78, 80
  bee larvae 32
  caterpillars 92
  carrion 132
  dead wood 26
  fungi 58
  grass 36, 154, 162
  honey 120, 124
  honeydew 132, 134, 138
  humus 160
  insects 34, 94, 96, 132, 134, 136, 144
  leaves 52, 54, 60, 138, 156, 160
  lichens and detritus 56, 58
  nectar 32, 78, 90, 92, 110, 112, 114, 116, 120, 122, 124, 132, 138
  plant sap/material 30, 34
  pollen 32, 90, 110, 112, 114, 116, 120, 122, 124
  seeds 136
  shoots and buds 50
  snow flies and mites 38
  spiders 28, 144
  stored foods 144
  water-borne organic particles 70
  wood 98, 158
Diptera **9**, 42, 64, 76
distribution
  Africa 76, 90, 98, 116, 160
  Asia 50, 54, 78, 80, 96, 98, 112, 114, 116, 122, 142
  Australia 60, 154, 158, 162
  Brazil 156
  Central America 138
  United States
    East and southeast 136
    South-central 36
    Southeastern 58
    Southwestern 32
  Europe 50, 54, 78, 80, 96, 98, 110, 112, 114, 116, 120, 122, 142
  Indo-Pacific 94, 134
  Mexico 138
  New Zealand 116
  North America 26, 30, 50, 52, 54, 70, 80, 90, 116
  Rocky Mountains 38
  South America 26, 90, 116
  worldwide 28, 34, 56, 76, 92, 132, 144
Dolophilodes distinctus 70
Dorylus spp. 130
dragonflies 8
Dunatothrips 60, 63

E
earwigs **8**, 16, 17, **34–5**
Embiodea **9**, 42, 56
Ephemeroptera 8
Ephydra hians 67
Entognatha 8
Eumeninae 92
Eumenes coarctatus 165
eusocial 20, 84, 92, 96, 104, 106, 120, 122, 124, 126, 147
Eustenogaster calyptodoma 83
evolutionary relationships 8–9

F
fall webworm 54–5
fleas 9
flesh flies 22
flies 9, **9**, 42
  alkali 67
  black 67
  crane 64
  fruit 67
  house 64
  true 64, 67
fontanella gun 158, 162
food webs 164–5
Forficula auricularia 34–5
Formicidae 126
froghoppers 18, **19**
fungus gardens **140–1**, **150–1**

funnels, cases, and stalk builders 64–81
    caddisfly cases 68–9
    case studies 70–1, 76–81
    materials and features 72–3

G

galleries **20–1**, 56, **57**, 84, **87**, **108–9**, 132, **133**, 136, **167**
glands
    abdomen 105, 110
    anal 19
    extrafloral 139
    genital 78
    head 133
    mouth 68
    salivary 45, 151
    silk 40, **42**, 80
    throat 106, **107**
glowworms **65**, **72–3**
gnats **65**, 67, 72
grasshoppers 8, 13, 16
Grylloblatta **38–9**
Grylloblattodea 16, 38
Gryllotalpa 36–7
Gryllotalpidae 16
grubs
    ambrosia beetle 20–1
    beetles 16, 17
    blister beetle 32
    carrion beetle 22, **23**
    dung beetle 24, **25**
    twig girdler 26

H

habitats
    bullhorn acacia plants 138
    desert 90
    domestic buildings 144
    fields 116

    forests 26, 30, 54, 58, 94, 98, 132, 134, 144, 160
    gardens 116
    grasslands 36, 78, 110, 112, 154, 160
    heathland 114
    meadows 80, 116, 120
    moors 114
    mountains 38
    rocks 56
    sandy ground 28, 32, 76, 114, 136
    savanna 124
    streams 70
    temperate 34
    tree trunks 56
    trees and shrubs 60
    tropical 34, 92, 94, 124, 134
    wetlands 144
    woodland 50, 52, 58, 78, 80, 96, 110, 112, 120, 142, 156, 158, 162
Habroscelimorpha dorsalis 28
Halictus sexcinctus **122–3**
heat, body 52, **53**, 120, 129
hemimetabolous 16, 147
Hemiptera **9**, 13, 16, 30
Hexapoda 8
Holometabola 9, 13, 16, 64
honeycomb structures 10, **104–5**, 166
honeydew 139, 142
hornets **15**, 82
    European **96–7**
    nests **15**
Hymenoptera 9, **9**, 42, 82, 90, 92, 94, 96, 98, 110–44
hyperparasites 142
Hyphantria cunea **54–5**

I

ice crawlers 16, **38–9**
iterative algorithms 10

K

kaydids 8
Kladothrips 62
Koptothrips 63

L

lacewings **9**, 42, 64, 67
    common green **80–1**
    pearly green **78–9**
Lasius fuliginosus 142
Lasius niger 128
Lasius umbratus 142
leaves **134–5**
leaf-cutter ants 10, 127, **140–1**
leaf insects 8, **12**
Lepidotera **9**, 40, 50, 52, 54, 67
lice 9, **9**
    bark 40
    book 58
life cycles
    caddisflies 67
    fall webworm 54
    rose leafroller 51
    wasp 86–7
Linepithema humile 144
longhorn beetles **26**, 27

M

Macrotermes bellicosus 146
Macrotermitinae 148, 150
Magicicada septendecim **30–1**
maggots 13, 22, **23**, 67
Malacosoma americanum **52–3**
mantids 42
mantises 8
Mantodea 8, 42
mayflies 8
Mecoptera 9
Megachile **118–19**

Melipona **124–5**
Melipona quadrifasciata **125**
Meliponini 102
Meloe franciscanus 32–3
metamorphosis 9, 13, **23**, 25, 30, 101, 119
mimicry 32, **33**, 49
mosquitoes 64, 67
moths **9**
   carpet 46
   case-bearing **47**
   clothes **46**, 47
   fungus 48
mud houses 92, **93**
mycangia 21
Myrmeleontidae 74–5
Myrmicinae 136

N
Nasutitermes triodiae 162
Nasutitermes walker 158
nests
   arboreal 158, 162
   carton 142, **143**, 146, 158, **159**
   galleries **20–1**, 56, **57**, 84, **87**, **108–9**, 132, **133**, 136, **167**
   underground 10, 11, 15, **86**, 110, **111**, 120, 122, 128, **129**, 142
nest material
   acacia spines 138
   animal bodies 32, 80
   bark 144
   feces 154, 158, 160, 162
   fungus 98
   grass 162
   ice 38
   leaves **12**, 34, 50, 60, 134, 144
   mud **7**, 30, 90, 92, 94
   paper 86
   plant fibers 116

plant resin 124
polyester, natural 114
rock 38
saliva 158
sand 28, 76
silk 52, 54, 56, 58, 60, 70, 78, 80, 134
soil 28, 34, 36, 110, 114, 122, 136, 144, 154, 156, 162
snail shells 112
snow 38
stone 144
twigs 26
wax 120, 124
wood fibers/pulp 94, 132, 142, 158
Neuroptera 42, 64, 78, 80
Neuropterida **9**
New Zealand
   fungus gnat **64**, **65**
   glowworm **72–3**
Notoptera 8
nymphs 13
   barklice 58
   beetles 16
   cicadas 30, 31
   earwig 34
   froghopper 19
   gall thrips 63
   mole crickets 36, 37
   scale insects 18
   termites 147
   web spinners 56

O
Odonata 8
Oecophylla longinoda 134
Oecophylla smaragdina 134
Oncideres 26
Orthoptera **8**, 36
Osmia bicolor **103**, **112–13**

P
Paraneoptera 9
pests **27**, 46–7, 50, 54,
Phasmatodea 8
pheromones 48, 97, 129
piracy 63
Plecoptera 8
Pogonomyrmex badius 136
Polistinae 92
pollination 166
Polyneoptera 8, 9
predators 28–9
Pseudomasaris vespoides **91**
Pseudomyrmex ferruginea 138
Pseudomyrmex spinicola 138
Pscoptera 40, 58
Psocodea 9, **9**
Psychidae **48–9**
Pterygota 8
pupal case 13, **41**, 43, 44, 67

R
replates 132
Richard Gilder Center for Science and Education **167**
rose leafroller **50–1**
royal jelly 106, **107**

S
sap 17, 19, 30, 140
scale bugs **18**, 19
Scarabaeoidea **24–5**
scorpionflies 9
security guards 138
silk 13, 14, 40–63, 166
   cocoons 40, **41**, 43, 44, **45**
   glands 40
   uses 43
silk moths 40, **41**, **42**, **44–5**
Silk Road 44

silkworms 40, **44–5**
Silphidae **22–3**
silverfish 8
Simuliidae 67
Siphonaptera 9
singing 30–1, 36
Sirex noctilio **98–9**
snow and ice 38, **39**
splittlebugs **19**
stick insects 8
stings 82, 85, 90, 94, 100, 116, 138
stoneflies 8
Strepsiptera 9
supercolonies 145
swarm architecture 166

T
temperature control 52, **53**, **129**, 136, 152–3
termites **9**, 146–67
    Caatinga **156–7**
    case studies 154–63
    cathedral 146, **162–3**
    Cubitermes **160–1**
    gardening fungus **150–1**
    magnetic **154–5**
    materials and features **152–3**
    mounds 10, **11**, **148–53**, 154, **155**
    mushrooms **151**
    tree **158–9**
    ventilation **152–3**
Termitomyces 150
thrips **9**, 42, 43
    acacia **60**, 61
    gall **62–3**
Thysanoptera **9**, 42, 60
tiger beetles **28–9**
Tinea pellionella **47**

Tineola bisselliella **46**, 47
traps 64, 73, **74–7**
trees
    acacia 43, **60**, **61**, 62
    birch 142
    bullhorn acacia **138–9**
    cherry 52
    deciduous 54, **55**
    eucalyptus 20
    fruit 134
    oak 89, 142
    pecan 27
    rotten 142, **143**
    willow 142
Trichoptera **9**, 42, 64, 67, 70
triungulins 32, **33**
tubular cases **47**, **69**
tunneling 16, 36, **37**, 38, **128**, 132, **137**, 146
turrets, mud 30, **31**
twig girdlers **24**, 25
twisty wings **9**

V
ventilation systems 10, **152–3**
Vespa crabro **96–7**
Vespidae 94–5
Vespinae 92

W
wasps 42, 82–101
    building materials 84–5
    case studies 90–9
    common **86–7**
    gall **88–9**
    hover 83, **94–5**
    jewel **100–1**
    life cycle 86–7

    materials and features 88–9
    paper 92
    parasitic **9**, 25, 28
    pollen **90–1**
    potter 83, **92–3**, **165**
    wood **98–9**
weaver ants 12
web spinners 42, **43**, **56**, **57**
web spinners and silk weavers **9**, 40–63
    silk from silkworms 44–5
    webbing and cases 46–7
    materials and features 48–9, 62–3
    case studies 50–61
weevils 20
whitefly 19
woodworms 20–1
wormlions **76–7**

X
Xylocopa spp. **108–9**

Y
yellowjackets 82, 92

Z
Zoraptera 8
Zygentoma 8

Acknowledgments and Picture Credits

AUTHOR ACKNOWLEDGMENTS

M.S.E. is grateful to N. Pierce and K. Crous for their patience and exceptional help through the composition and design of the work.

PICTURE CREDITS

t: top, b: bottom, l: left, r: right

The publisher would like to thank the following individuals and organizations for their kind permission to reproduce the images in this book. Every effort has been made to acknowledge the pictures, however, we apologize if there are any unintentional omissions:

123rf.com/golf609: 91

Alamy/FLPA: 12t; Petro Perutskyi: 14; piemags/nature: 27; blickwinkel: 29; blickwinkel/F. Hecker: 36; Minden Pictures: 39; blickwinkel: 43; Agencja Fotograficzna Caro: 45; blickwinkel: 48; Avalon/Picture Nature: 49b; piemags/nature: 50; StellaPhotography: 51; Vince F: 53; piemags/nature: 55; piemags/nature: 62t; Andrew Greaves: 64; fishHook Photography: 65; Nature Picture Library: 72; Nature Picture Library: 73b; blickwinkel: 74; blickwinkel: 9, 76, 77b; blickwinkel: 77t; Nigel Cattlin: 80; Vinícius Souza: 81; Stephanie Jackson—Australian wildlife collection: 83b; Hemis: 86; Nature Picture Library: 89b; ephotocorp: 90; Gillian Pullinger: 93; Ian Redding: 96; Andrew Darrington: 103tr; blickwinkel 103b; Nature Picture Library: 108; Peerasith Chaisanit: 109b; Rebecca Cole: 110; blickwinkel: 111; Christoph Bosch: 112; blickwinkel: 113t; Nick Upton: 113b; Nature Photographers Ltd: 114; blickwinkel: 116; Nick Upton: 118; BIOSPHOTO: 122; Bazzano Photography: 131t; Amnat Buakaew: 135; Grant Heilman Photography: 136; Patrick Lynch: 137tr; Juniors Bildarchiv GmbH: 146t; Images of Africa Photobank: 146b; Associated Press: 147; AfriPics.com: 151; Minden Pictures: 155b; Associated Press: 157; Minden Pictures: 161; WildPictures: 165

Alex Wild: 127t

Antón Vázquez Arias, 2024: 145t

Clemson University—USDA Cooperative Extension Slide Series: 26

David Andréen: 167b

Donald W. Hall, University of Florida: 59

Dreamstime/Ermess: 7

Dr Guido Bohne, Kassel: 123

ETH Zurich/Albert Krebs: 37

Gan Cheong Weei: 94

Getty/Paul Starosta: 7; Xvision: 9, 41; ullstein bild: 9, 78; HADI ZAHER: 11; imageBROKER/Siegfried Kuttig: 15; JimmyLung: 20; TEMUR ISMAILOV: 40; Abdul Latif: 49t; Michael Weidemann/500px: 52; DE AGOSTINI PICTURE LIBRARY: 79; FERNANDO MARRON: 125t; Tim Mason/500px: 126; PATSTOCK: 127b; Pavel Vorobiev: 131b; George D. Lepp: 144; Martin Harvey: 155t; imamember: 159

Goodfon.com/Alonso-brosmann: 73t

Hock Ping GUEK: 95

iNaturalist/Jiri Hodocek: 83t; Lucas Rubio: 124; https://www.inaturalist.org/observations/105316642: 143t

James Gilbert: 9, 60, 61

Jan Šobotník OIST: 160

Jason L. Robinson: 70

John B. Schneider: 9, 58

M. Zúbrik, NFC: 99b

Nathaniel Walton: 32

Nature Picture Library/Nature production: 17t; Jan Hamrsky: 66; Kim Taylor: 88; Rod Williams: 101; Solvin Zankl: 103tl; Dietmar Nill: 121; Mark Moffett: 145b

Nicky Bay: 77bl

Philipp Hoenle: 138, 142

Rawpixel.com: 28;

Science Photo Library/Ted Kinsman: 42; STEVE GSCHMEISSNER: 154

Shutterstock/sumroeng chinnapan: 2; Kathy Clark: 8, 30; Wirestock Creators: 8, 115 Ernie Cooper, 34; Filip Senigl: 9, 25; skippy666: 9, 56; RATT_ANARACH: 9, 158; Muhammad Fiqri Rahman: 10; Dragon Claws: 12b; Andries Combrinck: 16; Lukas Jonaitis: 17b; Tomasz Klejdysz: 21; Marek Velechovsky: 23; Mark Green: 31t; Neil Bromhall: 31b; Anest: 54; yod 67: 82; Henri Koskinen: 89m; MRS. NUCH SRIBUANOY: 92; Istvan Csak: 97; Aleksandr Rybalko: 102; Ihor Hvozdetskyi: 107; Celso Margraf: 125b; Artush: 132; Jay Ondreicka: 133; Deer worawut: 134; Michael Siluk: 140; meunierd: 152t; Hennadii Filchakov: 152b; Agarianna76: 153; 2021 Photography: 163

Thomas Eltz: 117t, 117b

Thrips-iD/Manfred R. Ulitzka: 62b; Laurence Mound: 63t; Stephen Bell: 63b

Vitor Corrêa Dias Gonçalves: 156;

Warren Photographic: 139

Wikimedia Commons/Sam Droege: 33; Nabokov: 35; Endemic Grylloblattid: 38; Luis Fernández García: 57; Hallvard Elven, Naturhistorisk museum, Universitetet i Oslo: 71t; Arnstein Staverløkk, Norsk institutt for naturforskning: 89t; Michaellbbecker: 98; Belchergb: 109t; Ivar Leidus: 120; Geoff Gallice: 130; CSIRO: 162; Zeete: 167t

William M. Ciesla: 99t

www.antstore.de: 143b